Lex Fullarton

Watts in the Desert

Pioneering Solar Farming in Australia's Outback

Lex Fullarton

WATTS IN THE DESERT

Pioneering Solar Farming in Australia's Outback

ibidem-Verlag
Stuttgart

Bibliographic information published by the Deutsche Nationalbibliothek
Die Deutsche Nationalbibliothek lists this publication in the Deutsche Nationalbibliografie;
detailed bibliographic data are available in the Internet at http://dnb.d-nb.de.

Bibliografische Information der Deutschen Nationalbibliothek
Die Deutsche Nationalbibliothek verzeichnet diese Publikation in der Deutschen
Nationalbibliografie; detaillierte bibliografische Daten sind im Internet über http://dnb.d-nb.de
abrufbar.

ISBN-13: 978-3-8382-0864-0

© *ibidem*-Verlag / *ibidem* Press
Stuttgart, Germany 2016

Printed in the United States of America

TABLE OF CONTENTS

ACKNOWLEDGEMENTS

This book acknowledges the Gnulli people who are the traditional owners of the land known to them as Mungullah. The town of Carnarvon and the Solex project on Grey's Plane are situated on Mungullah. We thank the Gnulli people for sharing this land with us.

A special thanks to my old friends Ron Crow and Bob Price who inspired the inception of this development.

Of course, to my wife Julie and my family, who made this possible and to my good friend John Craig who has acted as editor of this story—thank you.

To all of the *Fruitloops*, it is humbling to think that so many would think so much of a person who probably doesn't really deserve it. In particular I thank the influence of Tony and Oscar Sala. Without all of you this story would never have existed and certainly never prospered.

The Board and staff of Horizon Power, Western Australia's rural and remote energy utility, who were most supportive of the incorporation of renewable energy. At times conflict has arisen as to the role of the utility as a competitor in energy generation, but all that has resulted in active participation and full discussion of the difficulties faced by the changing face of energy technology. I believe that conflict has resulted in the outcome we have experienced. I thank those parties involved in these processes over the years. In particular Mr Brendon Hammond the Chairman of Horizon Power, and Mr Mike Laughton-Smith PSM for their personal dedication and time contributed to the rise of solar pv generation in rural and remote Western Australia.

To Jim and Wendy Andreoli for their valiant assistance in not only designing and constructing the Solex ice-works by also for their never failing support to keep it running efficiently and effectively.

A special thank you to the Rowe family of Caltex Starmart Carnarvon. Without their belief in the product, Solex solar ice would never have got to market. As with any new product there were teething problems are we refined bagging, sealing, curing and storage problems. Five years on we are masters at 'the ice game' but it was not always that way.

Finally acknowledgement is made to Genevieve Simpson who co-authored Chapter Five who incorporated her research into this book to provide social richness to its story.

ABBREVIATIONS

A	ampere
ABC	Australian Broadcasting Commission
ABARE	Australian Bureau of Agriculture and Resource Economics
ABRES	Australian Bureau of Agricultural and Resource Economics and Sciences
ABS	Australian Bureau of Statistics
ACT	Australian Capital Territory
ALP	Australian Labor Party
ASIC	Australian Securities & Investments Commission
ATO	Australian Taxation Office
BOM	Bureau of Meteorology
CEEM	The Centre for Energy and Environmental Markets (UNSW)
CER	Clean Energy Regulator
CSIRO	Commonwealth Scientific and Industrial Research Organisation
FCA	Federal Court of Australia (Single Judge)
FCAFC	Federal Court of Australia Full Court (3 or more Judges)
FiT	Feed-in Tariff
GST	Goods and Services Tax, see also VAT
GW	Gigawatt (1000Megawatts)
GWh	Gigawatt hour (1000 MWh)
Hz	SI unit for frequency hertz – 1 cycle per second
ITAA 1936	Income Tax Assessment Act 1936 (Cth)
kW	Kilowatt (1000 Watts)
kWh	Kilowatt hour (1000 Watt hours)
LGCs	Large-scale generation certificates, a type of REC
LP	Liberal Party of Australia
MPPT	Maximum power point tracking
MRET	Mandatory Renewable Energy Target
MW	Megawatt (1000 kW)
MWh	Megawatt hour (1000 kWh)
NASA	National Aeronautics and Space Administration (USA)
NEM	National Energy Market
NOCT	Nominal operating cell temperature
NWIS	North West Interconnected System
OECD	Organistaion for Economic Co-operation and Development
ORER	Office of the Renewable Energy Regulator (now CER)
pv	photovoltaic
REC	Renewable Energy Certificate see also LGC and STC
RET	Renewable Energy Target
RRPGP	Renewable Remote Power Generation Program
SEDO	Sustainable Energy Development Office (Government of Western Australia)
STC	Small-scale Technology Certificate, a type of REC; Standard Test Conditions for solar panel comparisions.
TBL	Triple Bottom Line
TAFE	Colleges of Technical and Further Education

UNSW	the University of New South Wales
USA	United States of America
UWA	the University of Western Australia
V	volt
W	watt
WA	Western Australia
WASC	Supreme Court of Western Australia

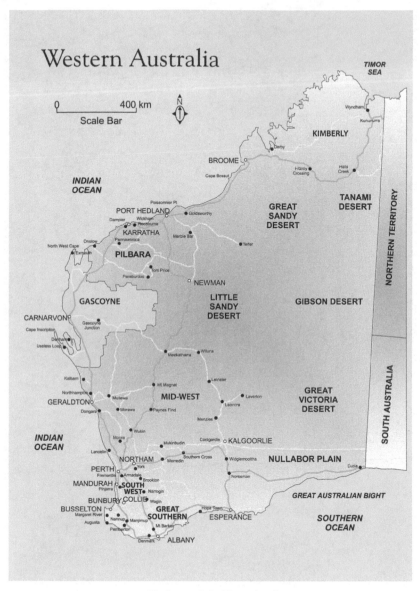

Western Australia

0 — 400 km
Scale Bar

N

INDIAN OCEAN

TIMOR SEA

Wyndham
Kununurra

KIMBERLY

Darby

BROOME
Cape Bossut
Fitzroy Crossing
Halls Creek

TANAMI DESERT

GREAT SANDY DESERT

Poissonnier Pt
PORT HEDLAND
Goldsworthy
Dampier Wickham
Roebourne
KARRATHA
Onslow Pannawonica
Marble Bar
Teifer

North West Cape
Exmouth

PILBARA
Tom Price
Paraburdoo

NEWMAN

GASCOYNE

LITTLE SANDY DESERT

GIBSON DESERT

CARNARVON
Cape Inscription
Gascoyne Junction
Denham
Useless Loop

Meekatharra
Wiluna

Kalbarri
Leinster
Northampton
Mt Magnet
Laverton
GERALDTON
Mullewa
MID-WEST
Leonora
Dongara
Morawa
Paynes Find
Menzies

GREAT VICTORIA DESERT

Wubin
Moora
Mukinbudin
Coolgardie
KALGOORLIE
Lancelin
NORTHAM
Southern Cross
Widgiemooltha
NULLABOR PLAIN
York
Merredin
PERTH
Fremantle
Armadale
Norseman
MANDURAH
Brookton
Pinjarra
SOUTH WEST
Narrogin
Euzla
BUNBURY
COLLIE
Wagin
BUSSELTON
GREAT SOUTHERN
Hope Town
Margaret River
Nannup
Manjimup
ESPERANCE
Augusta
Mt Barker
Pemberton
Denmark
ALBANY

GREAT AUSTRALIAN BIGHT

SOUTHERN OCEAN

INDIAN OCEAN

NORTHERN TERRITORY

SOUTH AUSTRALIA

The Gascoyne Region Western Australia.

LIST OF FIGURES

13

LIST OF COLOR PLATES

LIST OF TABLES

PREFACE

My story begins on the sailing ship *Minden*, anchored off the coast of Fremantle Western Australia, on the morning of October 14, 1851. Twelve-year-old John Fitzpatrick peered into the morning sun to see his new home in the burgeoning British Colony of Western Australia.[1] Thus began the story of the British Colonization of the Gascoyne Region of Western Australia and the part my ancestors played in it.

It is rather a long story and bears little relevance to the development of the Solex project, other than to provide context and background as to why I would launch into the hitherto little known world of harvesting solar energy using solar photovoltaic (pv) panels. The story of the Fitzpatricks' arrival in Western Australia, and how they trekked to the Gascoyne region in 1883, can be found in the book *Daurie Creek*, which was written by my cousin Merton Fitzpatrick in 2004.

Another cousin "Bonnie" Milne has written a number of books on the history of the development of the Gascoyne region and the parts played by the Fitzpatrick and Collins families in that history.[2] I am a great-great grandson of that boy who peered over the rail of the barque *Minden* and went on to establish pastoral stations in the Upper Gascoyne region of Western Australia, over 160 years ago.

For over 100 years, my family have raised sheep in the Gascoyne. Initially, the Gascoyne region was a "land of milk and honey" for the British Colonists, who brought their flocks of sheep to graze the virgin pastures of land that had served the Aboriginals who had wandered over and lived off the land for thousands of years. The original inhabitants maintained a semi-nomadic lifestyle and worked in symbiosis with the land, on which they depended for survival.

The British, on the other hand, ignorant of the sensitive nature of the natural environment in this near desert, simply saw it as an opportunity to exploit the apparently bountiful herbage to grow their wool. The wool provided by their sheep was sent by ship to feed the woollen mills of "Mother England." For nearly 100 years, the Glenburgh[3] wool rolled westward to the sea.[4]

However, by the 1960s, the sensitive desert scrub surrounding the town of Carnarvon had become a wind-blown, barren clay-pan. It was largely abandoned, desolate,

---

[1]   Merton George John Fitzpatrick *Daurie Creek* (2004) 4.

[2]   Grace Veronica 'Bonnie' Milne *Amelia* (2002); Grace Veronica 'Bonnie' Milne *Crossroads: A journey through the Upper Gascoyne* (2007); Grace Veronica 'Bonnie' Milne *Pioneer Father, Pioneer Son: York to Gascoyne with the Collins Family* (2010).

[3]   Glenburgh station is one of the Collins family's pastoral stations some 250 km east of the port of Carnarvon and situated in the central Gascoyne river catchment area.

[4]   Jack Sorenson, *The Lost Shanty* (1939) 16.

and useless to man or beast, except for the horticultural industry that clung to the banks of the Gascoyne River.

The land on which the Solex project was to develop had gone unnoticed, and unwanted, until it came to the attention of a great-great grandson of the boy who had peered over the rail of the barque *Minden* nearly 130 years earlier. That is the beginning of this story.

I was born in Carnarvon in 1956 and raised there, except for a short period in Broome, in the early 1960s. Mother's family were old pastoral station folk who loved a *wongi*,[5] usually over a very long cup of tea or better a beer or whisky (if there was any to hand). It was a very large family with links across the state and years. By the time I arrived on the Earth, most of the family members had moved out of the Bush to the City. However, whether they lived at Glenburgh, Carnarvon, or Mt Hawthorn (a suburb of Perth Western Australia), they were very much Bush people.

They all liked, or had liked, "a cool shandy on a hot day."[6] There was a sprinkling of "teetotallers,"[7] but in fairness, generally, that was more to do with health or economic reasons rather than moral temperance. Long and short though, they loved yarning and poetry about the days of old and the heroes of the Bush. The tales were sometimes a little far-fetched but "each of the tellers always swore that the tale that he told was true."[8]

From an early age, Father and I wandered the Bush around Carnarvon fishing and shooting. Our small family was not wealthy by any stretch of the imagination, but Mother and Father were very resilient and able to do quite a lot with very little. Mother was a devout Catholic, as were most of her family, and I was required to attend Mass regularly. Father considered the whole religious thing a confidence trick—he had little faith in the religious. It was under that parental influence that I grew up as an only child in Carnarvon.

As an only child, I was a little socially distant. To compound that isolation, most of the children in our neighborhood attended the State School while I went to St Mary's Catholic School. My school friends were Catholic and chiefly recent immigrants from southern Europe or "New Australians," as was the term in those days. The neighborhood chiefly comprised White Anglo-Saxon Protestants (WASPS). Both terms have fallen into archaic use and reflect a more pre-WWII era social culture that prevailed

---

5   The local Aboriginal word for a talk or chat.
6   A Bush term for a heavy drinker. A shandy is a mixture of beer and lemonade—before the advent of "light beers."
7   A person of temperate habit.
8   David Gordon Kirkpatrick (Slim Dusty) (Ballad), *The Frog* (1985).

in Carnarvon in the early 1960s. I developed a very multi-cultural attitude to society and find myself comfortable across a broad range of social environments.

Of particular significance to this story are two brothers who came to Carnarvon in 1963. They lived nearby, had Italian migrant parents, and also attended St Mary's. They were the sons of one of the scientists at the United States' National Aeronautics and Space Administration (NASA) space program at the Carnarvon Tracking Station, which was built in the early 1960s. We became lifelong friends and I was drawn into the excitement of the NASA space program.

To add to my "unusual upbringing," at the ripe old age of 12, I began spending my school holidays Jackerooing[9] on friends' sheep stations. On my 15th birthday, I commenced working as a bookmaker's clerk at the Exmouth racecourse, for an elderly bookmaker whose clerk had failed to turn up. It was decided that, despite my tender years, I had a Diploma in Bookkeeping from Newton's Business College and was studying bookkeeping at high school; so, for the Bush, in the early 1970s, that was ample enough qualification.

This background may appear a little irrelevant, but three important factors arise from that. I was learning to be resilient from living and working in the Bush, fencing, maintaining wells and windmills, and handling stock; learning the practical application of education; and, by making friends with the sons of rocket scientists, developing a thirst for experimentation. Those experiences were to later serve me well in conceiving and developing the Solex project.

---

[9]   An apprentice Pastoral Station manager.

CHAPTER 1—INTRODUCTION

"Two more long days from Rocky Pool, and then Carnarvon Town,
So sixty bales of Glenburgh Wool, from Inland heights go down."

John Alfred "Jack" Sorenson, Western Australian Poet (1907–1949)

1.1   INTRODUCTION

This book looks at the greatest challenge facing humankind. In the modern age, it is the destructive impact that the highest density of humans ever to exist on the planet is having on the Earth's ecosystem. It is not the planet that is in jeopardy, which is simply a mass of rock swirling in space; it is the carboniferous life forms that cling perilously to its skin that are in peril.

Alexander and Boyle state "society's current use of fossil and nuclear fuels has many adverse consequences. These include air pollution, acid rain, the depletion of natural resources and the dangers of nuclear radiation."[10] This book looks at the development of a renewable energy project, in rural and remote Australia, which addresses two of those problems—air pollution and the reliance on fossil fuel as an energy source.

The purpose of this book is to share the experience of the development of the first privately owned, commercial solar energy farm in Western Australia. Initially, it broadly examines the interaction between the, apparently competing, economic, environmental, and social factors that influence sustainable development. It then looks briefly at how the economic, environmental, and social influencing factors, referred to as the *triple bottom line* (TBL), have formed the philosophical viewpoints, which underpin Australia's principle political organizations—the "Socialist Left," the "Industrialist Right," and the "Environmentalist Greens." Broadly, the Left supports renewable energy, the Right opposes it, and the Greens promote it.

However, apart from consideration of Australia's attempt to reduce carbon dioxide emissions caused through fossil-fueled electricity generation, in-depth discussion of the views of the respective Australian political parties as to climate change generally is beyond the scope of this book. In 2001, Australia introduced legislation to establish a quantifiable target to displace fossil-fueled energy generation with energy sourced from renewable, and nonpolluting, sources such as wind and solar energies. The operation of Australia's renewable energy target (RET) is examined in this book as the

---

[10]   Gary Alexander and Godfrey Boyle, "Introducing Renewable Energy" in Godfrey Boyle (ed) *Renewable Energy: Power for a Sustainable Future* (2nd ed, 2004) 10.

sale of carbon credits created under that legislation forms part of the revenue stream of the Solex project.

The research supporting this book was conducted by way of comparing the identified and forecast economic, environmental, and social factors established at the commencement of the project, in 2005, with the findings of actual data collected during the period under review, and the economic, environmental, and social outcomes of the project in 2015. It also consists of a review of the legislation supporting Australia's RET. That examination also considers the financial impact on renewable energy and fossil fuelled energy producers.

It is acknowledged that the harvest of renewable energy is not a modern concept to Australia's Outback community. Outback pastoral properties have been using wind generators since the mid-1930s and the "Dunlite" wind generator is as iconic as the "Yankee" windmill used for pumping water. The reader is reminded that, for the first 10 years of British settlement in Australia, colonial settlements used no fossil fuel. Settlement began in 1788, and the first coal mining began in 1798.

The incorporation of solar pv panels into farm and pastoral homestead electricity generation was broadly taken up in the late 1970s and 1980s. However, while solar technology was being used on isolated properties, they were "standalone" systems exclusively for electricity supplies to the homesteads and surrounding buildings. As fossil fuel prices rose in the early 1970s, solar pv systems became essential to ameliorate the escalating costs of diesel fuel, as well as the costs associated with its transport and storage.

Of significance to the Solex project was the Australian government's policy of diverting fuel excises imposed on the sale of diesel fuel into a diesel fuel replacement program called the Renewable Remote Power Generation Program (RRPGP). As detailed in Chapter 3, funding from the RRPGP was sourced to finance the initial stages of the Solex project.

However, the subject of this book is not the small, standalone integrated energy systems common on Outback pastoral properties and farms, but solar pv electricity generation systems that began to be integrated into the distribution grids of national utilities in the 2000s. The use of dispersed, embedded solar pv installations configured as uninterrupted power supply systems is also beyond the scope of this book.

This book presents a case study of the solar energy project conducted over a period of 10 years from 2005 to 2015. The purpose of the project was to establish a solar energy installation to investigate the economic, environmental, and social benefits of renewable energy.

The principle objective of the Solex Carnarvon Solar Farm is to ameliorate environmental damage caused by over 100 years of industrial development in the Gascoyne region and to act as a catalyst to promote the benefits of renewable energy by the broader Australian community. It also intended to demonstrate renewable energy sources are not only practically achievable but also economically viable.

Discussed in Chapter 5, the community of Carnarvon were quick to adopt the new technology. On the basis of a self-help type of program where the people helped their neighbors, the solar community embraced the solar energy resource with great enthusiasm. The rate of acceptance of solar pv systems by the broader community in Carnarvon is illustrated in Figure 1.

Once termed "Fruitloops" by mainstream society, the solar community of Carnarvon realized the benefits harnessing the natural resource of the sun, and solar pv installations became normal household fixtures on roof tops across the town. It was not long before the electrical utility placed restrictions on further installations as it feared disturbances to electricity generation quality might negatively impact on the Carnarvon Town distribution system.

Citing potential stability issues effecting electricity supplies arising from a high capacity of dispersed embedded solar pv installations, the state utility placed a moratorium on the connection of solar pv systems in 2011. It is believed that Carnarvon was the first town in Australia to receive that distinction.

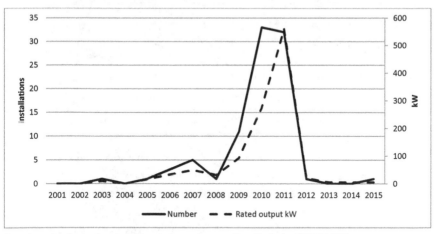

Figure 1:   Small-scale solar pv installation rates by year Carnarvon Town Distribution Grid: 2001–2015.
(Sources: Solex data, Australian Clean Energy Regulator, and Australian PV Institute[11])

---

[11]   Solex data held by Alexander Fullarton; Australian PV Institute, *PV Postcode Data* (2015) <http://pv-map.apvi.org.au/analyses> at July 22, 2015; Australian Government Clean Energy Regulator, *Postcode Data for Small-Scale Installations* (2015) <http://www.cleanenergyregulator.gov.au/RET/Forms-and-resources/Postcode-data-for-small-scale-installations> at July 22, 2015. Note there is a discrepancy between Solex data and the Clean Energy Regulator data as the CER data cover the entire postcode region

Note that the number of installations begins to rise from 2005 and, apart from 2008, continued until the state utility ceased to approve further connections in 2011. It was not until the end of 2012 when installations, approved prior to 2011, were completed.

The following Figure 2 illustrates the installation rate of solar pv systems in the broader Western Australian and Australian communities within the comparative time frame of 2001–2015. It indicates that a similar growth rate occurred in the broader community; however, that growth was a year later than in Carnarvon.

There are a number of factors that might contribute to that time delay, but a key factor was the source of financial support through government incentives. Part of those fiscal initiatives are examined in further detail in Chapter 2's review of Australia's RET legislation.

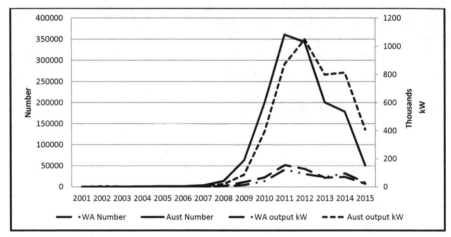

Figure 2:   Small-scale solar pv installation rates by year Western Australia and Australia: 2001–2015.
(Sources: Solex data, Australian Clean Energy Regulator, and Australian PV Institute[12])

The development of the Solex project was also to ascertain if solar energy could compete favorably in economic terms with fossil-fueled energy production. That demonstration was conducted in an industrial application, with the technology of solar pv installations available in Australia, in the early 2000s.

In 2005, the solar farm component of the Solex project was constructed, in Carnarvon Western Australia. Its purpose was to demonstrate that "a positive contribution to the *social and environmental* well-being of the community"[13] was not only, what

---

of 6701. That includes solar pv installations on pastoral properties for homesteads and water pumps. Solex data are for the grid-tie installations in the town of Carnarvon only. 2015 data to June 2015.

12   Solex data held by Alexander Fullarton; Ibid.
13   John Twidell and Tony Weir, *Renewable Energy Resources* (2nd ed, 2006) 2.

may have been considered by sections of the community as, at best an academic, or at worst an irrational, project development to combat atmospheric pollution and consequential climate change generally, but also be economically viable.

It has also formed a point of reference to attempt to unify the opposing political factions to adopt policies away from environmentally and socially damaging energy generation systems and encourage them to move toward developing enduring, or sustainable, industrial practices.

1.2   STRUCTURE

Initially, this book looks broadly at the concept of *sustainable development*. It examines the concept of the *TBL*, which considers the environmental and social consequences of economically, or fiscally, orientated development. It then focuses on the displacement of finite, and polluting, fossil fuel as a source of industrial energy by renewable, and nonpolluting, solar energy.

By harvesting non-polluting solar energy as an energy source, the Solex project demonstrates that an environmentally "sustainable development [can] be achieved whilst maintaining the ecological processes on which life depends"[14] and, at the same time, being profitable by giving a reasonable return on investment—that is, economically viable in fiscal terms. The social, and third, aspect of the Solex project is the creation of employment in occupations that were previously nonexistent—to contribute to the benefit to society.

It presents a description of how electricity generated from renewable energy sources contributes to combating climate change. It examines Australia's RET legislation to present an understanding of how the administrative processes of reducing atmospheric pollution from fossil-fueled energy sources operates. It describes how the Solex project was developed, the results of its operations, as well as its impact on social acceptance of solar energy and reduction of damage to the environment.

Chapter 2 explores the concept of the TBL and how its composite *economic, environmental, and social* factors interrelate. It suggests that the relationship is strongly influenced by overarching political forces, which attempt to place those influencing factors in balance with each other.

The chapter further suggests that rather than coordinating and balancing the TBL, in Australia, political interests often set the *economic, environmental, and social* factors in conflict and are destructive, rather than setting them in harmony for the benefit of current and future communities. It also examines the operation of the Australian Government's legislation supporting the RET. It looks at the fiscal impact of that

---

[14]   Ibid.

legislation on renewable energy producers and fossil-fueled energy generators and how the Solex project receives economic benefits from the creation and sale of "carbon credits."

Chapter 3 looks at the history, construction, and development of the Solex project. It looks at how the land was acquired and how the Solex project was conceived. It examines how the project developed to establish a research program from which reliable fiscal and scientific data could be collated from the solar pv installations.

In 2008, wind generation capacity was integrated into the solar enabling equipment to create a hybrid energy generation to the purely solar photovoltaic (pv) installation constructed in 2005. This innovation not only extended energy harvest beyond daylight hours but also provided data to monitor, compare, and contrast energy harvest effectiveness and efficiencies of solar and wind energy in Carnarvon.

Chapter 3 also describes how the Solex project "value added" to its renewable energy by way of constructing an ice-works powered from the solar/wind farm. The integration of the ice-works with the solar/wind farm was not only intended to improve the fiscal bottom line of the project but also to demonstrate a practical application of an "alternative use for alternative energy."

Chapter 4 examines the outcomes of the Solex project from the fiscal or *economic* perspective, as well as the *environmental* impact of "landcare" activities carried out on the previously barren land. It looks at the data collected from the Solex project and analyzes the economic and associated environmental outcomes from the aspect of avoided economic and environmental costs. It also looks at the outcomes of land restoration activities.

Chapter 5, contributed by Simpson, considers the broader *social* impact and outcomes of the Solex project and its influence on the attitudes of the general community and government organizations. It looks at how the example of the Solex project led to the community of Carnarvon adopting solar energy harvesting to improve their domestic, commercial, and industrial activities at reduced fiscal and environmental costs. It also considers the impact on the production of fossil-fueled based energy by the electricity utility servicing the region and the changing attitudes of that organization to solar energy integration in the period from commencement of the Solex project until the end of 2015.

Chapter 6 reviews and concludes the book. It brings the economic, environmental, and social findings and outcomes together. It outlines the limitations of the case study and suggests areas for further research and development. It suggests that the leadership displayed by the Solex project may have played a pivotal role in the acceptance of solar energy in broader community of Western Australia owing to its success.

1.3   SUMMARY

This chapter introduced the concepts to be examined in the book and laid out how it will address the issues relating to harmonizing the conflicting elements of the TBL. It looked at the growth of small-scale solar pv systems in Carnarvon, Western Australia, Australia. It laid out how the book is structured to provide an examination of Australia's RET and describes the experiences of the development of solar farming in Carnarvon, Western Australia, and the development of the Solex project in Carnarvon, Western Australia, over the decade 2005–2015.

Chapter 2 examines the concept of the TBL and how the groups of influencing factors relate to each other and the division between Australian political parties as they seek to represent their respective interests in relation to their philosophical position in the TBL structure. It begins by defining the TBL and then places the Australian political viewpoints in relation to their place in that structure. Finally, to provide context and background, it briefly examines an aspect of Australia's attempts to redress the problem of air pollution caused by the burning of fossil fuel—the Australian RET.

2.2   THE TRIPLE BOTTOM LINE

In 1987, the United Nations World Commission on Environment and Development issued its Report of the World Commission on Environment and Development: Our Common Future (the Brundtland Report). It defined sustainable development as being "development that meets the needs of the present without compromising the ability of future generations to meet their own needs."[15]

The report further defined sustainable development as

> a process of change in which the exploitation of resources, the direction of investments, the orientation of technological development; and institutional change are all in harmony and enhance both current and future potential to meet human needs and aspirations.[16]

From the concept of sustainable development, a corporate reporting system was developed, which incorporated the environmental and social impacts of its operations, as well as its fiscal accounting reports.

The *Macquarie Dictionary* defines the "triple bottom line (TBL)" as "a form of auditable company reporting which seeks to balance financial gain against responsibility to society and to the environment, in response to a corporate strategy that aims for economic, environmental and social gain."[17] Therefore, "progressive businesses"[18] or "responsible corporations" not only report their operations in terms of the fiscal, or economic, profit of an enterprise expressed in monetary units but also in terms of their contributions to the benefit of society generally, with consideration to the preservation and conservation of the natural environment.

Twidell and Weir suggest that "the aim of sustainable development is for the improvement [in the quality of life and standard of living of the world] to be achieved whilst maintaining the ecological processes on which life depends."[19] They further suggest that "at a local level, progressive businesses aim to report a positive *triple bottom line,* i.e. a positive contribution to the economic, social and environmental well-being of the community in which they operate."[20]

Strange and Bayley consider that "the core of sustainable development is the need to consider 'three pillars' *together*: society, the economy and the environment."[21] That framework for sustainable development has crystallized into the econo-enviro-

---

[15]   Gro Harlem Brundtland, *Report of the World Commission on Environment and Development: Our Common Future*, UN Doc a/42/427 (1987) 41.

[16]   Ibid 43.

[17]   Colin Yallop et al (eds), *Macquarie Concise Dictionary* (4th ed, 2006) 1311.

[18]   Twidell and Weir, above n 13.

[19]   Ibid.

[20]   Ibid.

[21]   Tracey Strange and Anne Bayley, *Sustainable Development: Linking Economy, Society and Environment* (2008) 27.

societal concept of the TBL. The influencing factors of economic–environmental–social impacts and the relationship between those factors are illustrated in Figure 3.

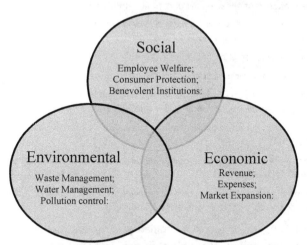

Figure 3:   An illustration of the triple bottom line factors incorporating some examples of each.

In 1990, the Australian Government recognized that ecologically sustainable development represented one of the greatest challenges to the nation's government, industry, and society in coming years. It recognized that there was no universally accepted definition of ecologically sustainable development but suggested the following definition:

> Using, conserving and enhancing the community's resources so that ecological processes, on which life depends, are maintained, and the total quality of life, now and in the future, can be increased.[22]

The national strategy document identified the goal for ecologically sustainable development as being "development that improves the total quality of life, both now and in the future, in a way that maintains the ecological processes on which life depends."[23] To support that strategy, the Australian Government enacted the Environment Protection and Biodiversity Conservation Act 1999 (Cth). The act is intended to provide a legal framework to protect and manage, nationally and internationally, important flora, fauna, ecological communities, and heritage places defined in the Act as matters of national environmental significance.

---

[22]   National Strategy for Ecologically Sustainable Development—Part 1 Introduction. Prepared by the Ecologically Sustainable Development Steering Committee, Endorsed by the Council of Australian Governments December, 1992, 1.
[23]   Ibid.

Specifically, section 3 (1) of the Act prescribes its goals:

(a) to provide for the protection of the environment, especially those aspects of the environment that are matters of national environmental significance; and

(b) to promote ecologically sustainable development through the conservation and ecologically sustainable use of natural resources; and

(c) to promote the conservation of biodiversity; and

(ca) to provide for the protection and conservation of heritage; and

(d) to promote a co-operative approach to the protection and management of the environment involving governments, the community, land-holders and indigenous peoples; and

(e) to assist in the co-operative implementation of Australia's international environmental responsibilities; and

(f) to recognise the role of indigenous people in the conservation and ecologically sustainable use of Australia's biodiversity; and

(g) to promote the use of indigenous peoples' knowledge of biodiversity with the involvement of, and in co-operation with, the owners of the knowledge.[24]

In September 2003, the Western Australian government produced a sustainability strategy policy that, among other objectives, was intended to

[d]evelop a State Strategic Planning Framework for the Public Sector that reflects sustainability and the triple bottom line [and]

[i]ncorporate sustainability principles and practices based on the Sustainability Act into relevant legislation as it is reviewed or drafted.[25]

In 2004, the Western Australian Local Government Act 1995 (WA) was amended to address a range of matters, including provisions to incorporate the sustainability themes into the content and intent of the legislation. The Act states:

In carrying out its functions a local government is to use its best endeavours to meet the needs of current and future generations through an integration of environmental protection, social advancement and economic prosperity.[26]

The Planning and Development Act 2005 (WA) introduced a specific purpose of the Act regarding sustainability.[27] Among the purposes of the Act, it specifically states that it is "to promote the sustainable use and development of land in the State."[28]

The emphasis on sustainability within the principal legislation governing planning practice in Western Australia is an important reflection of the role of promoting sustainable development through planning. In 2006, the City of Cockburn became one of

---

[24]   *Environment Protection and Biodiversity Conservation Act 1999 (Cth) s 3(1).*
[25]   Government of Western Australia, *Hope for the Future: The Western Australian State Sustainability Strategy: A Vision for Quality of Life in Western Australia* (2003) 55.
[26]   *Local Government Act 1995* (WA) s 1.3(3).
[27]   *Planning and Development Act 2005* (WA) s 3.
[28]   Ibid s 3.1(c).

the first local authorities in Australia, and the first in Western Australia, to adopt the definition of sustainability.

In 2011, the City of Cockburn adopted a sustainability strategy to embed that philosophy into its administration.

In the document prescribing its sustainability strategy, the City of Cockburn defines sustainability as:

> Pursuing governance excellence to meet the needs of current and future generations through an integration of environmental protection, social advancement and economic prosperity.[29]

Despite those changes to legislation and the adoption of environmental conservation policies by federal and local authorities, the *Sustainability Bill* (WA) was never introduced to the Western Australian Parliament. Therefore, the proposed *Sustainability Act* (WA) did not come into existence.

The brief history of the adoption of sustainability legislation in Western Australia indicates that the implementation of policies for sustainable development is determined by the actions of government.

While it is generally considered in Western Democracies that parliament reflects the "will of the people," in fact, in Australia, parliament tends to consist of a loose conflagration of political groups each with their own focus or political philosophy.

Fullarton was an Independent Candidate for the Gascoyne region's Western Australian parliamentary seat for three election campaigns: 2001–2008.[30] As such, he was involved in negotiations with the candidates of other political parties and closely studied the political philosophies of the major political factions in the Australian Parliaments.[31] He has noted the differing underpinning philosophies of each of those parties and their differing primary focus in relation to the primary factors of the TBL.

It is suggested that a fourth factor should be considered when examining the concept of the triple bottom line—that of its "binding agent"—the political philosophies of governing bodies, as illustrated in Figure 4.

---

[29]   City of Cockburn, *Sustainability Strategy 2013–2017* (2013), 5.
[30]   Western Australian parliamentary terms are for four years, unless otherwise terminated by the Governor of Western Australia.
[31]   Australia is a Federation consisting of one Federal Parliament and seven State and Territorial parliaments. The political parties of each state are broadly affiliated with, or part of, their Federal counterparts. The level of cohesion varies from party to party.

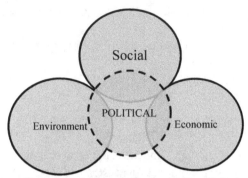

Figure 4:   The triple bottom line factors showing the political factor binding the relationship.

Figure 4 is an "ideal" framework. In Australia, the interests of political factions function in such fashion that, rather than government acting as a cohesive force, to co-ordinate and harmonize the TBL, or sustainability development framework, into a stable structure, the fragmented political factions tend to destabilize the harmony of the TBL.

In Western Australia, the "left-wing"[32] Australian Labor Party (ALP) government lost office in September 2008, and it appears that government action regarding 'sustainable development" ceased at that point. The succeeding "right-wing"[33] Liberal Party (LP) subsequently advised that the Western Australian Sustainability Strategy "is no longer being referred to as it was a policy of a previous government and is outdated."[34] It appears that governmental policies, and governmental approaches to the concept of sustainable development, differ between the "left" ALP and the "right" LP.

A detailed examination of left-wing and right-wing political philosophies is beyond the scope of this study; however, it is reasonable to suggest that the "left," generally represented by the ALP, tends toward social democracy and the welfare of society, while the "right," generally represented by the Liberal and National Parties, tends toward industrialism and the welfare of the economy in fiscal terms.

Furthermore, this book considers that, in Australia, it is the Australian Greens Party that represents environmental interests. However, they also state "today, the Greens recognise that speaking for the environment is not enough—we also need to speak on behalf of others who are disadvantaged in our society: children, refugees, students, individuals and families living in poverty."[35]

---

[32]   Yallop et al, above n 17, 684.
[33]   Ibid 1048.
[34]   Letter from Ken Baston Western Australian Minister for Agriculture and Food to Vince Catania Western Australian Member for North West Central, January 27, 2015 (copy held by author).
[35]   The Greens <http://greens.org.au/our-story> at February 13, 2015.

Figure 5 develops the illustration of the TBL shown in Figure 1 to show the Australian political parties' general philosophical perspectives.

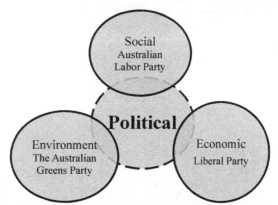

Figure 5:   The triple bottom line factors showing the Australian political parties representing the respective factors.

Further discussion in this book is based on the abovementioned broad assumption. It is suggested that, in attempting to bring all aspects of the TBL together, the Solex project is considered to be an amalgamation of the three political philosophies.

This book now moves on to discuss the impact of industrial development on the Earth's natural environment. In particular, it looks at the discharge of *greenhouse gases* into the atmosphere caused by the combustion of fossil fuels as an energy source. Those greenhouse gases affect the constituency of atmospheric gases and contribute to an increase in the Earth's atmospheric temperature, a concept referred to as *global warming*.

2.3   GLOBAL WARMING AND THE CARBON CYCLE

This section examines part of Australia's attempt to reduce the impact of *global warming*. While a detailed examination of the causes of *global warming* is beyond the scope of this study, there is good evidence that the major factor affecting global warming is attributed to a rise in the level of atmospheric carbon dioxide (CO_2). "We know for sure that, because of human activities especially the burning of fossil fuels, carbon dioxide in the atmosphere has been increasing over the past two hundred years and more substantially over the past fifty years."[36] It is generally accepted that action must be taken to mitigate the impact of rising CO_2 emissions on global warming.[37]

---

[36]   Sir John Theodore Houghton, *Global Warming: The Complete Briefing* (3rd ed, 2004) 8.
[37]   Ibid 10.

Prior to outlining the operations of Australia's RET, this section provides the following outline regarding mankind's impact on changing the carbon dioxide composition of Earth's atmosphere. The overview provides background as to why it is considered desirable to substitute fossil-fueled energy sources with nonpolluting, renewable energy generation systems.

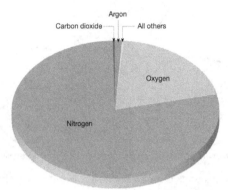

Figure 6: The modern composition of Earth's atmosphere.

As illustrated in Figure 6, the modern composition of Earth's atmosphere is roughly 78 per cent nitrogen and 21 per cent oxygen. The other 1 per cent is made up of 0.037 per cent carbon dioxide and the remainder, which is mostly inert gases such as argon, neon, and the like, as well as water vapor, methane, nitrous oxide, and chlorofluorocarbons.[38]

To provide an understanding of the dynamic processes "through which carbon is transferred in nature between [the atmosphere and] a number of natural carbon reservoirs,"[39] the illustrations in Figure 7 depict the carbon cycle.

---

[38]   Ibid 16.
[39]   Ibid 29.

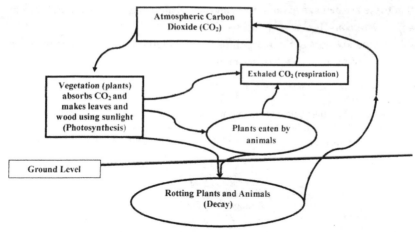

Figure 7:   The basic carbon cycle.

Houghton considers that

> [b]ecause carbon dioxide is a good absorber of heat radiation coming from the Earth's surface, increased carbon dioxide acts like a blanket over the surface, keeping it warmer than it would otherwise be. With the increased temperature the amount of water vapour in the atmosphere increases, providing more blanketing and causing it to be even warmer.[40]

While the concentration of carbon dioxide appears negligible, at less than 0.05 per cent of the total composition of the atmosphere, today, it is more than 25 per cent higher than at any time in the past 420,000 years. The recent increase in atmospheric CO_2 has occurred since the beginning of the "industrial" era (defined as since 1750), and most of that increase has occurred over the past 50 years. The increase in atmospheric CO_2 is primarily from the burning of fossil fuels (land-use changes and cement manufacturing also contribute), with half of this increase having occurred since the mid-1970s.[41]

To provide background and context, this book notes that, in order to indicate savings in fossil fuel and emissions, it is considered that 1 kWh (unit) of energy requires 300 mL of diesel fuel[42] and produces 650 mg of emissions in doing so.[43] It is pointed

---

[40]   Ibid 9.

[41]   Jonathan H. Sharp with assistance from Ferris Webster, John Wehmiller, Joseph Farrell, Willett Kempton, Ronald Ohrel and Douglas White "Increasing Atmospheric Carbon Dioxide" College of Marine & Earth Studies University of Delaware (2007) < http://co2.cms.udel.edu/Increasing_Atmospheric_CO 2.htm > at August 27, 2013.

[42]   Santiago Arnalich, *Epanet and Development. How to Calculate Water Networks by Computer* (2011) 153.

[43]   Soli J Arceivala and Shyam R Asolekar, *Environmental Studies: A Practitioner's Approach* (2012) 46. Some sources put this figure as high as 800 mg and as low as 600 mg. There are many influencing factors such as fuel efficiencies and mechanical and electrical power losses. For consistency, this book uses 650 mg

out that these are average assumptions, and actual data may vary considerably from system to system and fuel type to fuel type.

As illustrated in Figure 8, there are three general methods of addressing the issue of increasing atmospheric CO_2, i.e., by implementing:

- actions to reduce emissions caused by the combustion of fossil fuel by using substitute (renewable) fuels;
- actions to reduce the volume of existing atmospheric CO_2 by increasing physical carbon storage within the Earth's surface—carbon sequestration and increased vegetation; and
- actions to reduce energy consumption and, therefore, reduce demand for electricity generation.

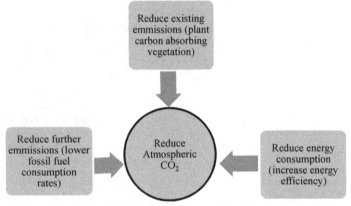

Figure 8:   Methods of reducing atmospheric CO_2

This book focuses on the first method—actions to reduce emissions caused by the combustion of fossil fuel by using substitute (renewable) fuels—and will discuss that method in the context of renewable energy produced by the Solex Carnarvon Solar Farm.

As part of its commitment to protect the environment by reducing atmospheric pollution, in particular, rising volumes of carbon dioxide caused by the combustion of fossil fuel, the Australian Government enacted legislation to encourage the generation of electricity from renewable sources.

The following section examines the legislation that supports Australia's legislative program to reduce atmospheric carbon dioxide emissions.

---

as referred to by Arceivala and Asolekar. That figure has been supported by David Harries, Adjunct Professor School of Electrical, Electronic and Computer Engineering, University of Western Australia.

2.4 AUSTRALIA'S RENEWABLE ENERGY TARGET

In June 2000, the Australian Government crystallized the desire to bring renewable energy into Australia's energy mix by passing a Bill to enact "an Act for the establishment and administration of a scheme to encourage additional electricity generation from renewable energy sources, and for related purposes."[44]

Undoubtedly enacting that legislation was in accordance with the Australian Government's response to the 1987 Brundtland Report and to meet its United Nations commitments through the respective enabling legislation—*the Environment Protection and Biodiversity Conservation Act 1999 (Cth)*.

The Bill was introduced to the Australian Parliament to "implement [sic] a commitment to introduce a mandatory target for the uptake of renewable energy in power supplies in order to contribute towards the reduction of Australia's greenhouse gas emissions."[45] The "mandatory target," initially referred to as the mandatory renewable energy target (MRET), became known more simply as the renewable energy target (RET).

The passing of the Bill created the *Renewable Energy (Electricity) Act 2000* (Cth) (REE Act). The basic principle of the Act is to "dilute" atmospheric pollution by requiring a purchaser of electricity to demonstrate that a percentage of that electricity has been generated from a renewable energy source.

In practical terms, the onus placed on each purchaser to "dilute" the consumption of fossil-fueled energy with renewable energy would be impossible to meet. The concept of physically "mixing" renewable energy-sourced electricity with fossil-fueled generated electricity at a central point is little more than a philosophical concept devoid of practicality. However, the RET scheme is somewhat ingenious in overcoming that problem. The Act places the onus of mixing renewable energy-generated electricity with fossil fueled-generated electricity on electricity wholesalers.

The administrative process by which the "dilution" is validated is rather simple. "A registered person may create a certificate for each whole Megawatt hour (MWH) of electricity generated by an accredited power station that the person operates during a year"[46] from an eligible renewable energy source. The "eligible renewable energy sources" are listed in section 17 of the REE Act.

To validate the "dilution," an electricity wholesaler must surrender a number of renewable energy certificates (RECs) for each 100 MWh purchased in each calendar

---

[44]   *Renewable Energy (Electricity) Act 2000* (Cth).
[45]   *Renewable Energy (Electricity) Bill 2000* (Cth) Explanatory Memorandum.
[46]   *Renewable Energy (Electricity) Act 2000* (Cth) s 18.

year commencing January 1, 2001.[47] That number percentage began at 0.24 per cent for the 2001 year and was originally intended to rise to 20 per cent in the year 2021.[48] At this point, it is sufficient to accept that a percentage is legislated. How the "dilution" is quantified is examined in greater detail in the following section.

Therefore, for every 100 MWh of electricity purchased by a wholesaler, that wholesaler produces a number of RECs, created from an accredited renewable power station. Should the prescribed percentage for a particular year be 5.98 per cent, as it was in 2010, then a wholesaler purchasing 43,800 MWh in a year must surrender 2,619 RECs to illustrate that 5.98 per cent of that electricity was generated from renewable energy sources. In that manner, the physical connection as to the precise energy mix of the source of supply becomes irrelevant on a broad scale.

An illustration of the process of "dilution" is presented in the following hypothetical example: A wholesaler, who purchases electricity in South Australia, which was purchased from coal-fired generators in the La Trobe Valley in Victoria, purchases RECs created by a solar farm in Western Australia from a trader in New South Wales to comply with the legislative requirements, surrendering those RECs to demonstrate the required "dilution" of emissions.

Alternatively, a wholesaler may choose to pay a "renewable energy shortfall charge" in lieu of surrendering a REC.[49] For convenience, the concept and economic consequences of the option of the shortfall charge are discussed at the end of this section.

To provide background, a brief history of the Australian RET is provided on the Australian Government's Clean Energy Regulator Internet website:

> Renewable energy has an important role to play in reducing Australia's greenhouse gas emissions and reaching the goal of 20 per cent renewable energy by 2020.

> Known previously as the Mandatory Renewable Energy Target, the Renewable Energy Target has been in operation since 2001, with the initial aim to source 2 per cent of the nation's electricity generation from renewable sources. In 2009, this was increased to ensure renewable energy made up the equivalent of 20 per cent of Australia's electricity (41,000 GWh).

> Since its beginning in 2001, the Renewable Energy Target has increased the number of installations of small-scale renewable energy systems, and successfully stimulated investment in renewable energy power stations. As at July 2014:

---

[47] Ibid s 4.
[48] Ibid s 39.
[49] Ibid s 36.

- Over 400 renewable energy power stations have been accredited, and
- Over two million small-scale renewable energy systems have been installed.

In 2013, over 20.4 million small-scale technology certificates and 14.6 million large-scale generation certificates were validly created.[50]

Discussed later in more detail, there were significant legislative changes to the REE Act in 2011. At this point, the most relevant change is to the concept of the terminology of the REC. In 2011, two classes of REC were created—small-scale technology certificates (STCs) for installations of less than 100 kW capacity and large-scale generation certificates (LGCs), which are created by installations in excess of 100 kW capacity. Therefore, for the purposes of this discussion, RECs created prior to 2011 are referred to as RECs, and those created after 2011 are referred to as either STCs or LGCs, as is appropriate. Where RECs have transitioned to LGCs, as in the case of the Solex Carnarvon Solar Farm, only the current term, LGC, is used.

By enacting the RET, the Howard Liberal Government had made a positive step toward sustainable development and reducing greenhouse gas emissions. Initially, the 2001 targeted reduction was a modest 0.24 per cent of Australia's electricity generation from renewable sources. However, the goal was to reach 20 per cent by 2020. This was to be a very worthy contribution to the reduction of atmospheric pollution by greenhouse gases.

While there has been significant scientific research and debate as to the precise volumes of atmospheric pollution caused by electricity generation, this book suggests that the RET is estimated according to broad assumptions. Finally, the actual percentages referred to are forecasts of estimates rather than quantifiable calculations.

The scope of inaccuracy created by the concept of forecasting, rather than the projection of statistically established data, has given rise to uncertainties. Those uncertainties have, in turn, given rise to overriding political debate as to the actual environmental and social implications of fossil-fueled electricity generation.

An examination of the data shown in Table 1 reveals the discrepancy between "the estimated amount of electricity that will be acquired under the relevant acquisitions during the year"[51] and the actual consumption for that year.

---

[50] Australian Government, Clean Energy Regulator, *The History of the Renewable Energy Target* (2014) Renewable Energy Target <http://ret.cleanenergyregulator.gov.au/About-the-scheme/History-of-the-RET> at March 29, 2015.
[51] *Renewable Energy (Electricity) Act 2000* (Cth) s 39.

Year	Required GWh of renewable source electricity[52]	Renewable power percentage[53] LGCs (RECs)	Estimated total electricity consumption (GWh); required GWH/RPP	Actual electricity consumption (GWh)[54]
2001	300	0.24	125,000	224,641
2002	1,100	0.62	177,420	227,563
2003	1,800	0.88	204,545	226,452
2004	2,600	1.25	208,000	236,581
2005	3,400	1.64	207,317	245,642
2006	4,500	2.17	207,373	247,568
2007	5,600	2.70	207,407	251,227
2008	6,800	3.14	216,560	257,998
2009	8,100	3.64	222,527	244,414
2010	12,500	5.98	209,030	241,586
2011	10,400	5.62	185,053	na
2012	16,763	9.15	183,202	na
2013	19,088	10.65	179,230	na
2014	16,950	9.87	171,732	na
2015	18,850	11.11	169,667	na
2016	21,431	tba		
2017	26,031	tba		
2018	28,637	tba		
2019	31,244	tba		
2020	33,850	tba		
2021–2030	33,000	tba		

Table 1: The renewable energy sourced electricity and the renewable power percentage charge. (tba, to be advised; na, not available)

To indicate the inaccuracies of "forecasting," Table 1 illustrates how the legislation establishes the relationship between the estimated volume of renewable energy-sourced electricity and the renewable power percentage charge required to achieve that outcome. It also projects the total volume of electricity estimated to be consumed to estimate the accuracy of the legislated percentage of LGCs required to be surrendered by liable parties.

Forecasting can only ever be a "best guess" of what is likely to happen, as there will always be factors, existing at the time of the forecast is given, about which the forecaster could never have been certain, or even aware.

Note that the RET, at the time of writing, was ultimately to be 33,000 GWh. Watts[55] suggests that it is only once an event has occurred that the circumstances leading to that event can be examined and rationalized. Therefore, this book argues that debate

---

[52] Ibid s 40.
[53] *Renewable Energy (Electricity) Regulations 2001* (Cth) reg 23.
[54] Australian Government, Department of Agriculture, Australian *Energy Statistics—Energy Update 2011* (2015) <http://www.agriculture.gov.au/abares/publications/display?url=http://143.188.17.20/anrdl/D AFFService/display.php?fid=pe_abares99010610_12c.xml> at June 13, 2015.
[55] Duncan J Watts, *Everything is Obvious: How Common Sense Fails* (2011) xiv.

surrounding the rigid application of strict quantitative data, and various energy production targets, distracts from the philosophical intent to "dilute" fossil fueled energy production with nonpolluting renewable energy sources. Furthermore, it is suggested that political influences will continue to alter the RET in the future.

It is suggested that the rigid adherence to what appears to be scientifically established or based data has merely supported political argumentation, which has become distracted from the prime objective—to replace polluting and limited energy resources with nonpolluting and unlimited energy resources.

However, Australia's use of solar energy is comparatively low when considered with the high rates of solar energy installations in other Organization for Economic Co-operation and Development (OECD) countries. This is despite the findings of an Australian Bureau of Agriculture and Resource Economics (ABARE) report that "Australia also has higher incident solar energy per unit land area than any other continent in the world."[56]

The report suggests that a primary influence in the exploitation of a country's solar resource is not the abundance of the resource, but rather, that "the distribution of solar energy use amongst countries reflects government policy settings that encourage its use, rather than resource availability."[57] It appears that, in Australia, political argumentation has become focused on the economic aspects and has become based purely on fiscal analysis. That has further distracted from the central argument of combating atmospheric pollution from the use of fossil fuel.

The RET has become a very sensitive political issue between the socialist favoring ALP, the economist LP, and the environmentalist Greens. In 2010, the Rudd ALP government increased the RET from its originally targeted 4.27 per cent for the 2010 year to 5.98 per cent and, further, to 10.65 per cent (more than half way to its original target of 20 per cent by 2020) in 2013.[58]

The ALP government also introduced other legislation aimed at promoting sustainable projects, such as its mining rental resource tax and a tax on carbon emissions. The Australian industrial corporations reacted strongly to those taxes, and a concerted political campaign, focused on repealing those taxes, was conducted by groups representing the political interest of those organizations, to replace the ALP government with a more economically biased Liberal government in the 2010 Federal election.

---

[56]   Geoscience Australia and Australian Bureau of Agriculture and Resource Economics, *Australian Energy Resource Assessment* (2010) 264.

[57]   Ibid 264.

[58]   Australian Government, Clean Energy Regulator, *The Renewable Power Percentage Target* (2014) Renewable Energy Target <http://ret.cleanenergyregulator.gov.au/About-the-Schemes/About-the-renewable-power-percentage/Annual-targets> at March 29, 2015.

In 2014, the Abbott Liberal government repealed both the mining tax and the carbon tax. Interestingly, the ultimate outcome of 20 per cent electricity generation from renewable sources by 2020 appears to have remained the RET for the Abbott Liberal government. However, it also appears that there was considerable political pressure to reduce, or even abolish, the RET for Australia in 2015.

To illustrate the political influence on the RET pricing mechanisms, Figure 9 presents an indication of how various political decisions have affected the LGC market prices during the period 2003–2011. This book suggests that the market price for LGCs became less of an influence on reducing pollution and became little short of a market place for speculation by "investors" manipulated by varying political philosophies, which prevailed from time to time in the Australian Parliament.

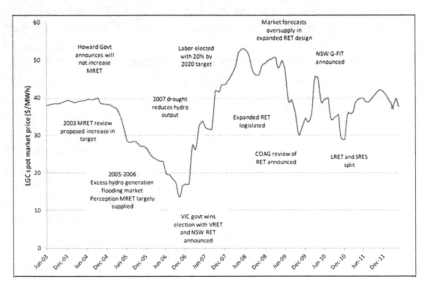

Figure 9:   Political influences on the LGC/REC market prices: 2003–2011.[59] (Source: ROAM Consulting Pty Ltd)

It is suggested that rather than attempt to accurately determine annual electricity consumption, which, at best, can only be forecasts or estimates, a subjective approach should be adopted to set the RET and the corresponding fossil-fueled energy tax rate.

To provide support for the concept of a subjective renewable power percentage for a desired annual RET, the following rationale is suggested. Data provided by the

---

[59]   ROAM Consulting Pty Ltd, "Solar Generation Australian Market Modelling" (Report to the Australian Solar Institute, 2012) 38. Underlying spot price curve sourced from Green Energy Markets and the Clean Energy Council.

Australian Energy Regulator as to the National Electricity Market (NEM) electricity consumption illustrated in Figure 10 suggests that Australia's total electricity consumption is somewhere in excess of roughly 200,000 GWh.

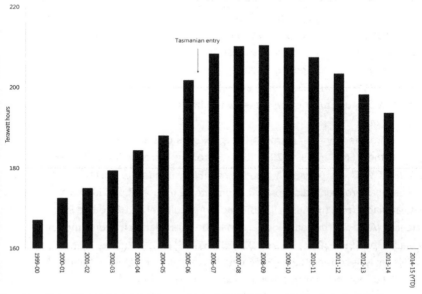

Figure 10: National Electricity Market electricity consumption: 2000–2014. (Source: Australian Energy Regulator[60])

As the NEM does not include electricity consumption in Western Australia or the Northern Territory, a further statistical basis encompassing those states is considered. Table 1 includes those states but is also subject to some underreporting. The data given in Table 1 suggest that the national average is closer to 240,000 GWh, though that data are some years out of date.

Be that as it may, if 200,000 GWh was accepted as a subjective baseline, then the RET "estimation" is simply a matter of deciding what percentage target is desired and the increment calculated from volume to percentage by dividing it by two. That is, a 20 per cent target is achieved by setting 40,000 GWh as the required GWh of renewable source electricity. Simplicity has long been held as one of the essential criteria of a "good tax system"[61] and is the basis for this recommendation; it may not, however, be politically desirable to some parties.

60   Australian Energy Regulator *National Electricity Market Electricity Consumption* (2015) <https://ww w.aer.gov.au/node/9765> at July 17, 2015.
61   CCH Australia Ltd, *The Asprey Report: An Analysis* (1975) 8.

2.5   THE IMPACT OF THE RET ON ENERGY GENERATORS

Australia's Renewable Energy (Electricity) Act

This discussion ignores the argument that fossil-fueled energy production has no effect on atmospheric pollution or "climate change." That philosophical position assumes that, given there is no impact, no RET is required. However, in order to consider that point of view, the "no impact" position requires no action is necessary to ameliorate the atmospheric pollution caused by combusting fossil fuel. Therefore, no "dilution" is required and accordingly all rates of "shortfall charges" and related calculations of fossil-fueled/renewable energy production, ratios, and fiscal impacts would be set at zero.

Data provided by the Australian Energy Regulator show that electricity consumption for the NEM[62] in 1999–2000, the year of the introduction of the REE Act, was 167,100 GWh.[63] As previously noted, the NEM is limited to eastern Australia, and, therefore, a broader indication of national consumption is provided from statistical data available from the Australian Bureau of Agricultural and Resource Economics and Sciences (ABRES). That data indicate that the total electricity consumption for Australia for the 1999–2000 year was 210,230 GWh.[64]

This analysis considers the ABRES data to examine section 40 of the REE Act, which prescribed that in order to attain 0.24 per cent of "dilution" for the 2001 year, 300 GWh of renewable-sourced electricity was required. ABRES statistics reveal that, in 2000–2001, Australian electrical consumption was 224,641 GWh. Therefore, the figure of 300 GWh was a mere 0.13 per cent of the total consumption.

Assuming that all electricity purchased was done so by "liable entities,"[65] then the volume of RECs required to be surrendered in 2001 would have been:

224,641,000 MWh × (0.24/100) = 539,138.

To consider the economic forces of supply and demand on price, the volume of RECs created in the 2001 year is examined. Clean Energy Regulator data show that 619,906 RECs were created in 2001, of the 1,708,454 that were eligible for creation. Therefore, in 2001, supply exceeded demand by around 15 per cent. Examined in detail in the

---

[62]   The Australian Energy Regulator (AER) regulates energy markets and networks under national energy market legislation and rules. Its functions mostly relate to energy markets in eastern and southern Australia. The National Electricity Market (NEM) covers Queensland, NSW, Victoria, South Australia, Tasmania, and the Australian Capital Territory (ACT). Australian Energy Regulator <https://www.aer.gov.au/> at June 12, 2015.

[63]   Australian Energy Regulator, *State of the Energy Market 2010*, (2010) 21 <https://www.aer.gov.au/> at June 12, 2015.

[64]   Australian Government, above n 50.

[65]   *Renewable Energy (Electricity) Act 2000* (Cth) s 35.

following section, it is expected that the trade price of an REC in 2001 would have been somewhat less than its theoretic value of the shortfall charge and its "tax effective" premium. It is noted that, as of July 2015, 655 RECs that were created in 2001 remained valid.[66]

The administrative position in 2010 remained relatively unchanged to that in 2001, save for the changes to the target volumes.

In 2010, the statistical position was:

Required GWh	12,500 GWh
Renewable Power Percentage	5.98 per cent
RECs Generated	41,008,102
Electricity Generated	241,586,000 MWh

Therefore, in 2010, the requirement of 12,500 GWh of electricity to be produced from renewable sources, as shown in Table 1, to produce a theoretically based RET percentage for that year of 5.98 per cent, had actually risen to only 5.17 per cent of the total electricity consumption in real terms and not the 5.98 per cent decided upon.

If it is assumed that all electricity purchased was done so by "liable entities,"[67] although, in fact, it is not, given the numerous exempt power generators in Australia, then the volume of RECs required to be surrendered in 2010 would have been:

$$241,586,000 \text{ MWh} \times (5.98/100) = 14,446,843.$$

Data provided by the Clean Energy Regulator shows that 35,524,421 RECs were created in 2010 of the 41,008,102 eligible for creation. As of July 2015, the "oversupply" of RECs continues, and at that time, 784,362 RECs from 2010 remained valid.[68]

A detailed annual analysis of compliance with the RET targets is beyond the scope of this book, and some 14 years later, the abovementioned calculation has the benefit of hindsight. In 2000, when the RET was established, the projected results were no more than a "best-guess" scenario. Section 39 of the REE Act provides that the annual percentage is established by regulation. Failing that a percentage is set by regulation, it outlines a formula to establish the renewable power percentage. It should be noted that, despite appearing to be a very precise formula, based on scientific principles, it is an "estimate."[69]

---

[66] Australian Government Clean Energy Regulator, *Register of Large-Scale Generation Certificates* <https://www.rec-registry.gov.au/rec-registry/app/public/lgc-register> at June 13, 2015.
[67] *Renewable Energy (Electricity) Act 2000* (Cth) s 35.
[68] Australian Government Clean Energy Regulator, *Register of Large-Scale Generation Certificates* <https://www.rec-registry.gov.au/rec-registry/app/public/lgc-register> at June 13, 2015.
[69] *Renewable Energy (Electricity) Act 2000* (Cth) s 39 (3) (b).

Despite the inaccuracies in the "dilution" calculations caused by forecasting models, and of course, resistance from "liable entities" to a new tax, the "dilution" system began to work. Liable entities also found a few "loop holes" in the legislation.

It is trite to state that, where taxes exist, tax evasion and avoidance will soon follow. The REE Act and its "shortfall charge"[70] are no more than the imposition of a carbon tax, no matter how well disguised. The government's role is a little different in this scenario, as the charge on "liable entities" is akin to a tax and the creation of RECs for sale is a form of a government subsidy. In this case, rather than flowing through "general revenue," the two parties (taxpayer and beneficiary) deal, more or less, directly.

The key "tax avoidance" matter is the "tax deductibility" of the "shortfall charge." Any charge made in accordance with the operation of the Act is not considered to be a "business expense" and, therefore, is not tax deductible.[71] On the other hand, if a "liable entity" purchases an REC from an accredited (renewable energy) power station, the cost of purchase of that certificate, to avoid the tax, is considered to be a business expense and, therefore, is tax deductible.[72] Liable entities, therefore, prefer to purchase RECs and claim the cost as a tax benefit, rather than pay the shortfall charge.

The accounting, or fiscal, implications between the payment of a nondeductible and the purchase of a deductible commodity raises the commercial value of an REC beyond the cost of the value of the shortfall charge of $40. By purchasing an REC, the liable party gains a commercial advantage of the amount of income tax that otherwise would have to be paid. In Australia, during the years under consideration, the corporate tax rate was 30 per cent.

A legal academic might refer to such a series of financial transactions as "acceptable tax avoidance," that is, the government has intentionally designed its legislation to operate in that fashion.

Therefore, with the introduction of the REE Act in 2000, the Australian government had effectively introduced a carbon tax and trading scheme, albeit that the accuracies of its RET were somewhat doubtful. That too created an industry for scientists, lobbyists, and commentators to investigate and debate the impacts of climate change, air pollution, the fairness of the tax regime, and so on.

For 10 years from its inception in 2001 until the end of 2010, the uptake of renewable energy and the carbon trading system established by the REE Act, its related legislation and regulations remained relatively unchanged. The modest liability of 1 REC

---

[70]   Ibid s 36.
[71]   Ibid s 7A.
[72]   Letter from the Deputy Commissioner of Taxation to Alexander Fullarton, January 5, 2006 (held by author) Notice of Private [Tax] Ruling 59756.

for every 400 MWh of electricity traded in 2001 increased incrementally to 6 RECs for every 100 MWh traded in 2010.

However, Australia's reasonably effective and efficient "carbon trading system" became complex and convoluted in 2011. The Rudd ALP government amended the REE Act to encourage the installation of small renewable energy generation units (SGUs). The amendments were targeted at encouraging households to install primarily solar pv and solar hot water systems in their homes. The discussion of solar hot water systems is beyond the scope of this book other than to note that such systems are also eligible to earn RECs.

The 2011 legislative amendments effectively added a second tier of RECs to a "liable party's" requirement to surrender RECs in accordance with the volume of electricity purchased.

The REE Act was "amended" as to the creation and surrender of RECs. Division 4—creation of certificates, which consisted of 11 sections contained within four pages of legislation, was increased to 22 sections contained in 14 pages. The Act expanded from 102 pages in 2001 to 231 by 2015.

It is suggested that tax legislation in Australia generally has evolved into an extremely complex and convoluted legal discipline. Examining the fascinating history of Australia's income tax legislation is beyond the scope of this book; however, s 160 of the *Income Tax Assessment Act 1936* (Cth) (ITAA 1936) is referred to as an example of the ever-developing complexity of Australian taxing legislation.

The once single section, which had, at one time, related to a rebate in the case of the disposal of the assets of a business of primary production, by 1998—after an attempt was made by the Keating ALP government to "reform" the legislation by at least renumbering the ITAA 1936, through its Tax Law Improvement Project (TLIP)[73] —had expanded to more than 63 entire Divisions of over 450 sections ranging from s 160 AAAA to s 160 ZZZJ.[74] Similarly, s 23 of the REE Act had expanded into 10 sections and ranged from s 23 to s 23F.

Furthermore, in 1997, an attempt to reform Australia's income tax legislation produced the outcome of a second income tax assessment act working in conjunction with the ITAA 1936. The *Income Tax Assessment Act 1997* (Cth) (ITAA 1997) was intended to entirely replace the ITAA 1936, but the process was truncated by a change

---

[73] The Keating ALP government attempted to reform the income tax legislation by introducing the *Income Tax Assessment Act 1997*. The *ITAA 1997* was to have eventually replaced the *ITAA 1936* over a suggested period of five years. Some changes did occur but the ALP lost the government in 2000 and the opposing interests of the Liberal Party abandoned the Tax Law Improvement Project (TLIP) in favor of its A New Tax System (ANTS) series of tax legislation. Australia currently operates with two pieces of income tax legislation: the *ITAA 1936* and the *ITAA 1997*.

[74] *Income Tax Assessment Act 1936* (Cth).

of government, and now, Australia has two concurrent income tax assessment acts. Given that the REE Act is merely 15 years old, compared to the ITAA 1936 of nearly 80 years, it will be interesting to see how far the complexity of the REE Act 2000 develops over the forthcoming 65 years.

In 2011, the Rudd ALP government adopted a policy of encouraging householders to install domestic solar pv systems. The intent was to create increased subsidies, funded by liable parties' to reduce the capital cost of primarily domestic installations.

The amendments to the REE Act created an additional class of REC—the STC. The creation and required per cent of that system of REC creation trade and the surrender of those certificates was conceived to run in parallel with the RECs, which were renamed as LGCs.

This book does not examine the STC system in depth, as its operations are complex and beyond the scope of the Solex project, which is an "accredited" power station and, therefore, not considered a "small generation unit." This is despite the Solex Carnarvon Solar Farm being less than 100 kW capacity in 2011.[75] However, to provide the reader with background and context, the following discussion is given to illustrate the role of political philosophy in enacting legislation aimed at encouraging renewable energy sources to reduce atmospheric pollution caused through the use of fossil-fueled energy-based electricity generation.

An RET was not set for the STCs system, but a renewable power percentage was regulated. Regulation 23A of the *Renewable Energy (Electricity) Regulations 2001*, as amended in July 2015, set the small-scale technology percentage as following:

(a) for 2011—14.80%;
(b) for 2012—23.96%;
(c) for 2013—19.70%;
(d) for 2014—10.48%;
(e) for 2015—11.71%.

For comparison, the renewable power percentage for LGCs for the same years, as shown in Table 1, is as following:

(k) for 2011—5.62%;
(l) for 2012—9.15%;
(m) for 2013—10.65%;
(n) for 2014—9.87%;
(o) for 2015—11.11%.

The amendments of 2011, targeted at encouraging small domestic solar pv installations and solar hot water systems, are complex. Liable parties are required to comply

---

[75]　Regulation 3 (2) (c) defines a *small generation unit* as being 'a device whose energy source is solar (photovoltaic) if: (i) it has a kW rating of no more than 100 kW; and (ii) it generates no more than 250 MWh of electricity each year.

with both renewable power percentages. Therefore, they must purchase and surrender both LGCs and STCs to comply with the provisions of the REE Act in order to avoid the shortfall, which was increased to $65 as part of the amendments.

As of January 2011, the total liability for "liable parties" is, therefore, the combined cost of the purchase and surrender of LGCs **AND** STCs.[76] Thus, in 2012, for every 100 MWh of energy purchased by a liable party, that party was required to surrender 9.15 LGCs AND 23.96 STCs. The total cost to the liable party in 2012, assuming it had purchased those certificates from a registered renewable energy generator in January 2012, would have been:

9 LGCs @ $40.00	=	$360.00
24 STCs @ $32.50	=	$780.00
Total purchase	=	$1,140.00[77]

In terms of the basic unit of the energy market, the kilowatt hour (kWh), the impost was $1,140.00/100,000 kWh or 1.14 cents/kWh. The effect of the income tax benefit/liability to both parties is discussed in the following section.

However, the fiscal advantage of purchasing the tax deductible credits over paying the nontax deductible shortfall charge is obvious:

33 × $65 = $2,145 (2.145 c/kWh)

Accounting entries for the creation and surrender of RECs is reasonably simple. Fullarton has adopted a primary production style of "livestock account."

Given that livestock increase naturally and are purchased, sold, and die, the accounting treatment of the creation, trade, and surrender is considered by Fullarton to be similar. The only exception is that RECs cannot be killed for internal use (rations). However, if a liable party, a fossil-fueled generator, was to construct a solar or other renewable energy generation system to supplement its energy demands, then it could surrender its own registered RECs and fulfill the conditions that would create that type of transaction.

The consideration of creating RECs in a similar way to breeding sheep or cattle supports the argument that solar energy harvesting is nearer to the concept of the primary producing activity of farming than it is to the industrial activity of fossil-fueled energy production through the use of internal combustion engines or similar mechanical devices.

---

[76] Email from Timothy Huntly Gordon, Liable Party Accountant, to Alexander Robert Fullarton, June 29, 2015.

[77] Prices sourced from Solex accounting records.

In the accounts of Solex, a "Renewable Energy Certificate Account" contains similar terminology as found in a primary producer's Livestock Trading Account, *mutatis mutandis*. Of significance to this book is the definition of a "primary producer" under Australian taxation law. It is noted that, in Australia, a primary producer is not considered such under the ITAA 1997 simply due to the taxpayer "carrying on a primary production business"; the business must be defined as a "primary production business" pursuant to section 995-1.

A detailed examination of political influence on tax legislation and administration in Australia is beyond the scope of this book; however, a brief illustration of the definition of primary production, as it applies to solar farming, is given to provide context.

The hunting of animals is not considered a primary production business, as it lacks the consideration of animal husbandry. In Australia, a kangaroo shooter, who hunts the outback for kangaroos for meat and pelts, is subject to the vagrancies of the natural environment for his "harvest." He is not considered to be a primary producer. On the other hand, a fisherman, who hunts in the sea for fish and crustaceans for meat and other products, who is also subject to the vagrancies of the natural environment and who does not husband his livestock, is considered to be a primary producer. The difference is that fishing is defined as a primary production business in s 995-1 and hunting is not.

A cereal farmer who uses solar energy to grow his crops and harvests the product is considered to be a primary producer in s 995-1, but at the time of writing this book, a solar farmer who harvests solar energy to produce electricity is not. Therefore, tax concessions provided to primary producers as defined in s 995-1 are not afforded to solar or wind farmers. This is despite the similarities of harvesting their products from the physical environment and their production being subject to the vagrancies of the weather patterns of that environment.

There are other such convolutions in other Australian taxation legislation such as Australia's Goods and Services Tax (GST); however, a detailed discussion of Australia's taxation legislation per se is beyond the scope of this book.

Australia's Carbon Tax 2011–2014

However, it is noted that, in addition to the RET legislation that is considered in this section, during the period 2011–2014, Australia had a *Clean Energy Act 2011* (Cth) under which a carbon tax of $23 per tonne of carbon dioxide emitted (a carbon unit) was levied on industries that emitted in excess of 25,000 tonnes of pollutant gases per annum.[78] Though a detailed discussion of that legislation is beyond the scope of

---

[78]   *Clean Energy Act 2011* (Cth) s 20(4).

this book, it is useful to briefly examine that legislation, as the tax impacted the market for RECs and the costs of operation to liable parties under the RET.

The legislation defined a *carbon unit* and defined the cost in 2012 to be $23, and it rose each year to $25.40 by 2014.[79] It appears that the carbon unit was to apply in a similar fashion to the shortfall charge of the REE Act and applied to CO_2 emissions, rather than to energy produced by a generator.

The legislation was extremely complex, with many exemptions and offsets for certain enterprises, such as "high emitters" and "trade sensitive" industries. Also, how the emissions were to be calculated was not explained in simple terms. Regulation 8.3(5), quoted below, prescribed the method by which emissions, expressed in tonnes of CO_2 (equivalent), was to be calculated.

Regulation 8.3(5)

Step 1
Work out the emissions (E_{ij}), in CO_2-e tonnes, of each greenhouse gas (j) released by the operation of the facility during the relevant period from the combustion of each fuel (i) consumed by the facility for the purpose of producing electricity, as follows:

$$E_{ij} = \frac{Q_i \times EC_i \times EF_{ij}}{1,000}$$

where:

Q_i is the quantity of the fuel (i) consumed by the facility for the purpose of producing electricity as reported for the facility under subparagraph 4.22 (1) (a) (i) of the NGER regulations for the relevant period.

EC_i is the energy content factor of the fuel (i) as reported for the facility under paragraph 4.07 (2) (a) or 3 (b) of the NGER regulations for the relevant period.

EF_{ij} is the emissions factor determined as follows:

(a) if Method 2, 3 or 4 was used for reporting the fuel (i) and gas (j) in relation to the facility under the NGER Act for the relevant period— the factor reported for the facility under paragraph 4.07 (3) (a) of the NGER regulations for the relevant period;

(b) in any other case—the factor specified in Schedule 1 to the *National Greenhouse and Energy Reporting (Measurement) Determination 2008* for the relevant period.

Step 2
Add together the E_{ij} amounts worked out for the facility under step 1.[80]

79 *Clean Energy Act 2011* (Cth) ss 94–100.
80 *Clean Energy Regulations* 2011 (Cth) reg 8.3(5).

The carbon units were then surrendered at a cost of $23 each for every tonne of CO_2 emitted.

A model of how the carbon tax was to impact the Australian economy was carried out in 2011 by Siriwardana, Meng and McNeill. The analysis suggested that, with a carbon tax rate of $23 per tonne:

> Australia's real GDP may decline by 0.68 per cent, consumer prices may rise by 0.75 per cent, and the price of electricity may increase by about 26 per cent as a result of the tax. Nevertheless it allows Australia to make a substantial cut in its CO_2 emissions. The simulation results imply an emission reduction of about 12 per cent in its first year of operation. The tax burden is unequally distributed among different household groups with low-income households carrying a relatively higher burden.[81]

Perhaps, a rate of 1.495 cents per kilowatt hour is a closer true cost of a carbon tax rate of $23 per tonne, if the generally accepted standard that 300 mL of diesel and 650 mg of greenhouse gas emissions, per 1 kWh of electricity,[82] is applied as follows:

> 1,000 L = 1,000 kWh = 650 kg CO_2 or 0.65 tonne × $23 = $14.95/1,000 kWh = 1.495 c/kWh.

Many states implemented charges to electricity prices to comply with the carbon tax, but there appears to be little basis to establish a carbon tax rate of 2.1868 c/kWh in Western Australia.[83]

In late 2013, the Dairy Australia Limited commissioned an analysis of Australian Dairy Shed energy costs, which

> revealed some indicators suggesting a "rack" carbon price rate of $0.022 to $0.0235/kWh in Victoria, $0.02368/kWh in Western Australia, and $0.00375/kWh in Tasmania (reflecting that State's reliance on hydro-electricity). If passed on in full, this translated to $0.80-$6.40 a day, or 1.5–13% of total bills. However, the extent to which power companies passed on the carbon price in full or in part is unclear.[84]

Ultimately, political factors focusing on the economic costs of the carbon tax, and perhaps its inherent complexity, which clearly leads to uncertainty in compliance requirements, resulted in the carbon tax being repealed in 2014.

The LP of Australia ran an electoral campaign in 2013 that focused, in part, on repealing the carbon tax. By the end of 2015, the Prime Minister remained focused on

---

[81] Mahinda Siriwardana, Sam Meng, and Judith Mc Neill, "The Impact of a Carbon Tax on the Australian Economy: Results from a CGE Model" (Working Paper No 2011—2, School of Business, Economics and Public Policy Faculty of the Professions, University of New England, 2011) 4.

[82] This standard assumption will be examined further in Chapter 4.

[83] The costs of the carbon tax levied by the Western Australian electricity utilities from 2011 to 2014, as indicated by the reduction in the electricity tariff levied by Horizon Power to Solex, September 1, 2014.

[84] Dairy Australia *Australian Dairy Shed Energy Costs* (2015) <http://www.dairyaustralia.com.au/~/media/Documents/Environment%20and%20Resources/22072014-Australian%20Dairy%20Shed%20Energy%20Costs-Fact%20Sheet-July14.pdf> at August 8, 2015.

the success of the government in repealing the carbon tax legislation. Australia continues to have a carbon tax in the form of the RET and its accompanying legislation. According to the aforementioned formula and calculations, in 2015, that cost was around $52 per LGC and $40 per STC (or 1.046 c/kWh or $17.43 per tonne).

The graph illustrated in Figure 11 shows that Australia's carbon dioxide emissions peaked around 2005, continued to decline until around 2014, and have begun to rise since then. They are projected to be above the 2005 levels by around 2017. It appears that, despite the successes of the RET and Carbon Tax Legislation in reducing Australia's carbon dioxide emissions, since the repeal of the Carbon Tax, Australia's emissions have begun to rise once more.[85] There may be other economic, environmental, and social factors influencing that rise; however, the softening of the RET and the repeal of the carbon tax appear to be a strong influence on the reversal of the decline.

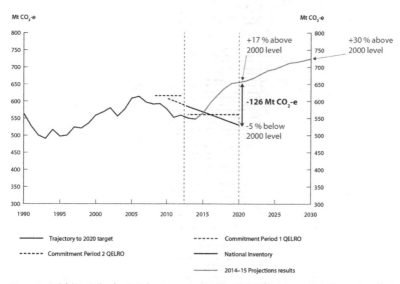

Figure 11: Australia's historical and projected emissions, along with the 2019–2020 abatement task. (Source: Australian Government, Department of the Environment, and Australia's Emissions Projections 2014–2015)

It is suggested that the introduction of the carbon tax, in addition to the rising compliance costs imposed by the RET legislation, distorted the *economic, environmental, and social* balance of the TBL in Australia and that economic interests influenced Australia's political landscape to "restore" the balance. In 2013, economic interests prevailed over the environmental and social influencing factors of the TBL in the minds of Australian electors/taxpayers.

---

[85]   Australian Government, Department of the Environment, *Australia's Emissions Projections 2014–15* (2015).

By the middle of 2015, environmental and social interests were opposing the economically driven interests of the Australian political landscape, and the RET remained a key focus of political groups, party policies, and electoral platforms at that time.

To place Australia's anti-atmospheric pollution measures in a global context, in 2010, Australia had one of the highest greenhouse gas emissions rates in the world. Figure 12 shows that, at per capita rate of 24.48 tonnes per person, Australia had the highest rate of the entire OECD member countries and over double the average of 11.69 tonnes per person of its fellow members.

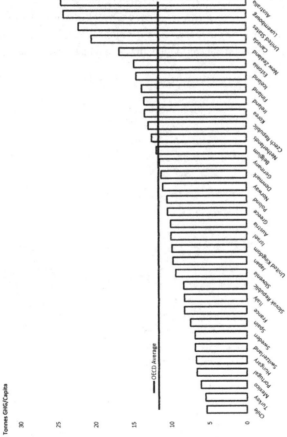

Figure 12: OECD national greenhouse gas emission intensities per capita, 2010. (Source: OECD Environment Statistics (database), 2012[86])

86   Organisation for Economic Co-Operation and Development, *Environment; Air and Climate; Greenhouse Gas Emissions by Source* (2016) < http://stats.oecd.org/Index.aspx?DataSetCode%3DAIR_GHG> at January 9, 2016.

As to amelioration of greenhouse gas emissions through renewable electricity generation, Table 2 shows that Australia placed 18th on the United States of America's Energy Information Administration's list of OECD countries' data base of renewable electricity generation volumes for 2012.

Rank	Country	2008	2009	2010	2011	2012
	OECD	1781.398	1834.263	1935.897	2073.911	2156.166
1	United States	392.736	429.652	440.231	527.490	508.360
2	Canada	384.587	379.476	366.247	398.286	397.344
3	Germany	93.980	99.251	109.635	126.780	142.685
4	Norway	139.051	125.309	116.988	120.883	142.412
5	Japan	106.065	106.833	115.238	117.692	122.368
6	Sweden	81.603	79.894	82.642	83.747	96.967
7	Italy	59.445	70.537	78.600	85.325	91.804
8	Spain	62.691	74.577	97.908	87.280	86.757
9	France	75.052	71.032	79.693	66.294	82.776
10	Turkey	34.165	37.869	55.319	57.686	64.637
11	Austria	44.775	47.352	45.162	42.473	50.881
12	Mexico	48.590	36.512	47.327	46.336	43.857
13	United Kingdom	23.292	26.902	27.124	36.254	40.248
14	Switzerland	38.119	37.813	38.246	34.702	40.155
15	New Zealand	27.939	30.919	32.599	33.539	31.564
16	Finland	27.764	21.780	24.263	23.395	28.233
17	Chile	27.072	29.396	24.079	24.537	25.219
18	Australia	19.610	19.620	22.357	26.813	23.834
19	Portugal	14.860	18.507	28.520	24.257	19.314
20	Iceland	16.342	16.709	16.931	17.083	17.423
21	Poland	6.792	8.892	11.104	12.906	16.868
22	Denmark	10.951	10.820	13.151	14.682	14.674
23	Netherlands	10.995	12.346	12.763	14.026	12.063
24	Belgium	5.478	6.749	7.788	9.665	10.473
25	Greece	5.736	8.150	10.576	8.215	10.107
26	Czech Republic	3.721	4.639	5.900	7.218	8.016
27	Korea, South	4.427	4.751	6.341	7.536	7.123
28	Slovakia	4.542	4.883	5.911	4.827	5.420
29	Ireland	3.578	4.099	3.725	5.422	5.229
30	Slovenia	4.271	4.862	4.702	3.846	4.283
31	Hungary	2.469	3.010	3.170	2.770	2.644
32	Estonia	0.197	0.541	1.044	1.181	1.477
33	Israel	0.025	0.099	0.147	0.313	0.475
34	Luxembourg	0.322	0.307	0.312	0.302	0.327
35	Puerto Rico	0.156	0.174	0.153	0.149	0.148

Table 2:   OECD countries volumes of renewable electricity generation: 2008–2012.
(Source: US Energy Information Administration)

Australia's 23,834 GWh compares with Chile's at 25,219 GWh and Portugal's at 19,314 GWh but is far behind that of the United States (508,360 GWh), Canada (397,344 GWh), Germany (142,685 GWh), Norway (142,412 GWh), and Japan (122,368 GWh).[87]

It is important to note that these data are for total renewable energy generation, which includes hydroelectric generation and other systems that are beyond the scope of this book. In addition, not all renewable energy generators are subject to Australia's renewable energy legislation, as they may have been in operation prior to 2000. Therefore, a comparison of the data shown in Table 2 with the data in Table 1, or other data in this book, may not reconcile. This comparison is to indicate Australia's global position in renewable energy generation only.

It is significant to differentiate between rates of greenhouse gas emissions as illustrated in Figure 12 and renewable energy generation volumes as shown in Table 2. Australia's renewable energy generation volume may be reasonably high, at nearly 24,000 GWH, as shown in Table 2; however, as it also has the highest rate of emissions, at nearly twice that of its fellow members of the OECD, it remains a net polluter. Therefore, in order to reverse the impact of greenhouse gas emissions, Australia needs not only to increase its renewable energy generation but also to decrease its per capita emission intensity.

A considerable force in addressing those factors is the perspective of the government of Australia in considering the influencing factors of the TBL. In 2015, the perspective of the LP of Australia, as to the relationship between the environmental and economic influences of the TBL, was indicated in an address given by the Prime Minister to the South Australian Liberal party annual meeting.[88] Mr Abbott said:

> When it comes to emissions per capita, our target—a target that we are absolutely confident that we can and will meet—is the best in the world. So, let's not have anyone say that this is a Government which is indifferent to environmental outcomes. This Government cares passionately about the environment. We only have one planet. We must leave it in better shape for our children and our grandchildren but the last thing we should ever do is clobber the economy to protect the environment because if we clobber the economy, the environment will surely suffer.

This statement appears to indicate that the LP places economic interests above environmental interests when considering greenhouse gas emission reduction measures. That perspective is consistent with the representation illustrated earlier in Figure 5.

---

[87] US Energy Information Administration *International Energy Statistics* (2012) <http://www.eia.gov/cf apps/ipdbproject/iedin-dex3.cfm?tid=6&pid=29&aid=12&cid=CG5,&syid=2008&eyid=2012&unit=BKWH> at August 29, 2015.

[88] Anthony John Abbott, "Address to the South Australian Liberal Party Annual General Meeting" (Speech delivered at the South Australian Liberal Party Annual General Meeting, Adelaide, August 15, 2015).

In conclusion, Australia's approach to reducing atmospheric pollution, through the implementation of the RET system, is essentially an enforceable limit of industrial carbon dioxide emissions through tracing the issue, trade, and surrender of registrable certificates—a "cap and trade system." A detailed discussion of that concept of combating pollution is beyond the scope of this book.

The solar farm provides Solex with a renewable energy source that competes directly with fossil-fueled reliant industries. That renewable energy source provides a production and marketing advantage to the Solex project. This discussion describes how the Solex project is an economic beneficiary of the Australian RET scheme.

This section has looked at the principles and the basic operation of Australia's RET in terms of volumes of the creation and consumption of carbon credits. To provide clarity about the impact of the RET system in fiscal terms, the following section examines the operation of the RET carbon trading system in monetary terms.

2.6   ACCOUNTING FOR THE RET

The preceding section examined the structure of Australia's RET system from a philosophical perspective—the reason for the legislation and the principles of its operation. In that examination, similarities with taxation legislation were considered. The *ITAA 1936* was compared to the RET and carbon tax legislation to illustrate the economic and social impacts of the differing political philosophies of governments on legislation.

To provide clarity for the reader, the fiscal examination of the treatment of accounting entries in the financial reports of liable parties is provided here, inclusive of the income tax implications of the RET.

A tax is defined as "a compulsory monetary contribution demanded by a government for its support."[89] There is no compulsion on a government to apply tax revenue in any particular manner. Revenue raised from tobacco taxes may be spent on railway infrastructure or parliamentary travel. Alternatively, the tax revenue may be allocated to expenditures on cancer research or medical and hospital facilities.

Similarly, the previously examined carbon tax was purely to the benefit of government revenue. It was not directly allocated to renewable energy infrastructure, but according to the socially based policies of the ALP government of the day, some of the revenue was allocated to the reduction of the income tax of lower income earning taxpayers and to social welfare benefits for pensioners. There was no compulsion for the government to match that revenue with expenditures on renewable energy infrastructure.

---

[89]   Yallop et al, above n 17, 1255.

However, an alternative legislative approach to the RET legislation may have been by way of a taxation system in which liable parties could have been levied with a "pollution" tax and the revenue raised allocated to subsidies, grants, or rebates paid to renewable energy producers.

The RET system imposes a liability on fossil-fueled energy generators based on the amount of energy they produce. It is, effectively, a consumption tax. The purchase of RECs in order to comply with the RET requirements, in lieu of a nontax-deductible fine or penalty, is a system of tax avoidance.

Therefore, this book considers that the imposition of purchasing and surrendering RECs, by way of a composite purchase of LGCs and STCs, by a liable party is reasonably considered to be a tax on those liable parties. The revenue received by renewable energy producers is similar to a rebate, subsidy, or grant from the government. In that manner, the RET system permits a direct transfer of revenue from the liable parties to renewable energy producers around the taxation system.

This section now looks at the accounting treatment of those fiscal transactions and the income tax implications thereof.

In order to examine the relationship of nontax-deductible expenses to tax-deductible expenses in revenue terms, the following formula is provided.

Value of nontax-deductible penalty = p
Value of tax-deductible REC = r
Tax rate = t
$p/(1 - t) = r$

Prior to 2011, the "shortfall charge" was set at $40[90]; the Australian corporate tax rate prevailing in the period 2001–2010 was 30 per cent, and, therefore, the value of an REC, in accounting terms, to account for the effect of taxation created before the tax market value of an REC, was:

$$\$40/(1 - 0.3) = \$57.14$$

In 2011, the shortfall charge was raised to $65[91] for both LGCs and STCs; the corporate tax rate remained unchanged at 30 per cent, and, therefore, the before tax market value rose to:

$$\$65/(1 - 0.3) = \$92.86$$

In fact, as illustrated in Figure 9, due to the free market forces of supply and demand, the traded price of RECs never reached those levels.

---

[90]   *Renewable Energy (Electricity) (Charge) Act 2000* (Cth) s 6.
[91]   *Renewable Energy (Electricity) (Large-scale Generation Shortfall Charge) Act 2000* (Cth) s 6; *Renewable Energy (Electricity) (Small-scale Technology Shortfall Charge) Act 2010* (Cth) s 6.

The sale of RECs/LGCs created from harvesting renewable solar energy forms part of Solex's revenue. The following extract of Solex's accounts illustrates its method of accounting for the creation and sale of RECs/LGCs.

Extract of Solex accounts 30th June 2010
Renewable Energy Credits (Livestock A/c)

		Number	Value	
Sales				$0.00
less	Opening stock	34	$1,467.72	
	Natural increase	182		
	Purchases	157	$7,065.00	
		373	$8,532.72	
less	Closing Stock	373	$15,812.72	−$7,280.00

Gross Profit from RECs   **$7,280.00**

A liable party's accounts do not reflect similar accounting treatment, as the entity does not "breed" or manufacture RECs. The accounts of the liable party disclose the purchase of RECs as a revenue item in its expense account. Alternatively, the disclosure of the shortfall charge is disclosed as a taxation expense in its profit distribution account, as it does with other taxes, dividends, and the like.

2.7   SUMMARY

This chapter has examined the concept of sustainable development and the need to consider its "three pillars"—the economic, environmental, and social impacts—as one interrelated concept: the *TBL*. The rationale of the TBL is that no single influence can exist in isolation for an indeterminate period of time. An imbalance between the three overarching factors in favor of one over the interests of the other two will adversely affect the others and ultimately itself. An enterprise focused solely on profit, at the expense of society and the environment, will not be able to be sustained indefinitely.

The chapter focused on the impact of carbon emissions caused by industrialization and its reliance on fossil-fueled energy sources. It briefly examined how the carbon cycle functions and successive Australian governments' legislative attempts to address carbon emissions to mitigate the impact of greenhouse gases on global warming.

Finally, it placed the impact of those legislative requirements on industry to dilute carbon emissions with energy sourced from renewable sources in the context of how the Solex project receives economic benefits from that legislative intervention. It also notes the distinction between national rates of greenhouse gas emissions on a per capita basis and the total annual greenhouse gases emitted by nations.

Australia has a relatively low annual volume of greenhouse gas emissions and a moderate renewable energy electrical generation program when compared to its fellow OECD member nations. However, it also has the highest per capita emission rate of the OECD member nations, at nearly double the OECD average.

The relatively low emission volumes is disguised by its sparsely populated nation, which covers an area comparable to the entire United States of America, with less than 10 per cent of the US population, at around 22 million persons. This gives rise to the belief that Australia does not have a greenhouse gas emission problem. In 2015, the attitude of the Australian Liberal government toward environmental conservation is being distorted by preferences toward the economic considerations of the TBL.

The physical environment in which the Solex project is located, and how it was developed, is discussed in the following chapter. Chapter 3 provides background and context to demonstrate how renewable energy can effectively compete in an industrial application, which was previously the realm of fossil fuel—an alternative use for alternative energy pioneered in Australia's Outback.

CHAPTER 3—THE SOLEX PROJECT

"Good, better, best. Never let it rest.
Until your good is better and your better is best."

St Jerome, Christian Saint and Scholar (347–420)

3.1   INTRODUCTION

This chapter looks at the implementation of the Solex project. It begins by examining the natural environment in which the project is located. In particular, it looks at the climate of the Gascoyne region, its rainfall patterns and atmospheric temperatures. That examination provides an understanding of the weather patterns that give rise to the setting of the desert landscape of the arid regions of Western Australia.

It then looks at the history of the land on which the Solex project is located and how the acquisition of that land created a demand for an economically sustainable project to justify the resources expended in acquiring the land. The arid landscape and the desert climate encouraged industrial development in an innovative manner—the Solex Carnarvon Solar Farm and ice-works.

It was the nature of the environment that caused the land to be abandoned shortly after European settlement. It was that natural environment that influenced an alternative economic land use, which exploited the very factors that caused it to be abandoned—boundless solar energy in a waterless environment.

In turn, the development of the Solex project had an ongoing social influence as to the creation of employment and the reduction of energy costs to commerce and industry.

3.2   LOCATION AND CLIMATE

The Solex project is located at Carnarvon, in Western Australia, depicted in Figure 13. It is situated just below the Tropic of Capricorn, some 1000 km north of Perth, on the coast where the desert meets the sea. Carnarvon is considered a desert climate as the Bureau of Meteorology (BOM) statistics reveal that it receives an average of just over 200 mm (8 inches) per annum.[92]

In addition to the primary aims, it is believed that, by creating a "waterless" farming alternative, Solex aims to inspire commerce and industry to investigate businesses

---

[92]   Australian Government Bureau of Meteorology, *Climate Statistics for Australian locations: Carnarvon Airport* (2014) < http://www.bom.gov.au/climate/averages/tables/cw_006011.shtml> at January 3, 2016.

and production activities, which do not rely on the vast consumptions of water and fossil fuels.

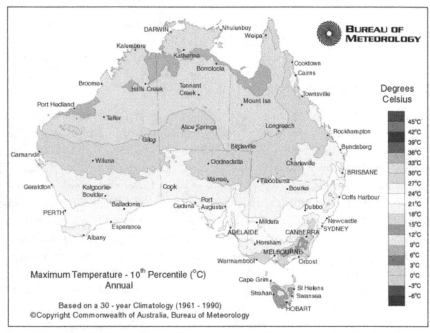

Figure 13: Map depicting climate and location of Carnarvon Western Australia. Note: the original map is in color. Isotherms are all above 9°C. (Source: Australian Government Bureau of Meteorology[93])

Local folklore subscribes to the concept that it never rains in Carnarvon. Valli told the tale about a newcomer to Carnarvon who

> was being initiated at the [Carnarvon] Club into the story about Carnarvon's nickname, Elsewhere. (The weather forecasts often conclude with the phrase "Fine elsewhere").
>
> "Has it always been like this?" he asked. "As far back as I can remember." replied the old-timer. "Why, even in the Bible; remember when it rained for forty days and forty nights? Well even then Carnarvon only got ten bloody points."[94]

However, as with many urban myths, it is not quite true. The graph illustrated in Figure 14 shows BOM statistics over the past 70 years. It reveals that Carnarvon receives an average of 226 mm (8 inches) per annum.[95] The rainfall is not zero, but that is certainly near enough for its climate to be classified as desert.

---

93 Australian Government Bureau of Meteorology <http://www.bom.gov.au/climate/map/temperature/IDCJCM0005 temperature.shtml> at July 19, 2007.
94 A point is about 2.5 mm of rainfall; Jack Valli *Gascoyne Days* (1983) 112.
95 Australian Government Bureau of Meteorology, above n 92.

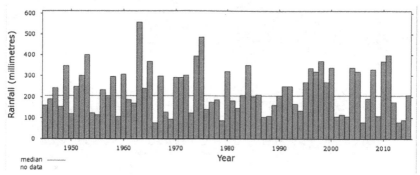

Figure 14: Carnarvon Airport annual rainfall: 1945–2015. (Source: Australian Government Bureau of Meteorology and Climate Data Online[96])

The Macquarie Dictionary defines a desert as "an area so deficient in moisture as to support only a sparse, widely spaced vegetation, or none at all."[97] Other sources, which focus on climate, such as the BOM and Western Australian Department of Agriculture, quantify the definition by suggesting that deserts are defined as area of low rainfall.

BOM suggests that regions with less than 350 mm of annual precipitation are reasonably considered a desert climate. Its website states "For example, the climate maps on our website use a condition of median annual rainfall less than 350 mm to describe an arid area.[98] The Western Australian Department of Agriculture, considers regions of less than 250 mm to be 'arid'."[99]

To provide an understanding of the Gascoyne Region in broader context, the average annual rainfall of Gascoyne Junction, some 156 km east of Carnarvon, is illustrated in Figure 15. It is observed that the average annual rainfall for Gascoyne Junction is slightly lower at 215.5 mm.

However, it is also noted BOM statistics for that weather station is available for a longer period than those for Carnarvon and cover 108 years, from 1907 to 2015. Therefore, the statistical information may depict a more accurate evaluation as it is based on a greater range of data.

---

96   Australian Government Bureau of Meteorology, above n 92.
97   Yallop et al (eds), above n 17, 323.
98   Australian Government Bureau of Meteorology *Glossary* <http://www.bom.gov.au/lam/glossary/dpa gegl.shtml> at January 6, 2015.
99   Andrew A Mitchell and David G Wilcox, *Plants of the Arid Shrublands of Western Australia* (1988) 10.

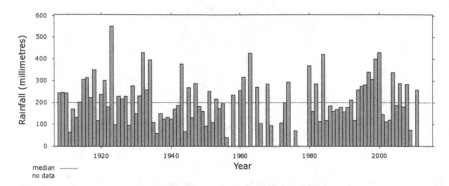

Figure 15:   Gascoyne Junction annual rainfall: 1908–2015. (Source: Australian Government Bureau of Meteorology and Climate Data Online[100])

Carnarvon is contained in a climate region that Sturman and Tapper describe the climate as dry arid with a warm temperature.[101] They look at traditional classifications of regional climate in Australia and state the 1865 South Australian example that considered the 300 mm (12 inches) annual isohyet as a safe boundary to grow wheat.[102]

However, as well as considering rainfall as a factor determining regional climate, they point to the energy balance[103]—a ratio between solar radiation and the ability of the ground to dissipate that energy (the Bowen ratio).

In 1996, Sturman and Tapper stated:

> In Australia there is concern about the serious problem of dryland degradation or desertification, defined as "land degradation in arid, semi-arid and dry sub-humid areas resulting from various factors including climatic variations and human activities" (at the United Nations Conference on Environment and Development). In Australia, overgrazing and associated vegetation removal, have an impact on dry rangeland areas, exacerbated at times of drought. Overgrazing and other forms of human-induced desertification affect the local and regional climate.[104]

The interaction of grazing animals within the carbon cycle is illustrated in Figure 7, in Chapter 2. Some researchers do not believe that overgrazing by livestock is a contributing factor to desertification. Savory considers that grazing of grasslands is an essen-

---

[100]   Australian Government Bureau of Meteorology, above n 92.
[101]   Andrew P Sturman and Nigel J Tapper, *The Weather and Climate of Australia and New Zealand* (1996) 300.
[102]   Ibid 298.
[103]   Ibid 306.
[104]   Ibid 327.

tial element in maintaining good soil conditions to improve carbon storage. He considers that grazing of livestock prevents conditions that give rise to desertification, which can occur despite high rainfall events in some regions.[105]

On the other hand, Mitchell and Wilcox, who also consider the role of vegetation in preventing wind and water erosion, suggest that while plant litter is essential in giving soil the ability to absorb water easily, "growth not consumed is not wasted, but is actually necessary for the maintenance of good range condition."[106]

The Gascoyne river catchment area, and in particular Carnarvon, has been subjected to those desertification impacts on its climate. However, the focus of this book is on the reduction of greenhouse gas emissions from burning fossil fuel and not the processes of desertification.

Of significance to this book, and discussed later in Chapter 4 under the subheading physical environmental impacts, is a pattern of flood events in the Gascoyne River delta. Long periods of drought followed by high rainfall events result in significantly increased water shed by the desert soils enhanced inability to absorb water. The Gascoyne River over tops its banks and flood waters extend across the delta flooding the town of Carnarvon generally and the Solex project specifically.

Those flood events pose a significant threat to Carnarvon and its horticultural region but, as explained in Chapter 4, can have positive outcomes for the rehabilitation of the land containing the Solex project.

Those events appear to be decadal, and examination of Figure 14 tends to support the hypothesis that they occur at the beginning of each decade when higher than average rainfall has been preceded by long periods of drought. According to Western Australian Department of Water publications, Carnarvon experienced significant floods in 1951, 1960, 1961, 1980, 1995, 2000,[107] and 2010.

While agriculture generally, and the pastoral industry in particular, has suffered greatly from negative impacts as a result of climate change, some of those changes have been positive influences for solar energy generation.

Boyle points to the advantage of clear skies as an economic benefit to large solar [pv] power plants.[108] Carnarvon, with its annual average of 211 clear days, has be-

---

[105] Allan Savory and Jody Butterfield, *Holistic Management: A New Framework for Decision Making* (2nd ed 1999).

[106] Mitchell and Wilcox, above n 99, 19.

[107] Government of Western Australia, Department of Water, *Water Facts WF13; Flooding in Western Australia* (2000) <https://www.water.wa.gov.au/__data/assets/pdf_file/0014/1670/WER-120-WRCWF13.pdf> at January 3, 2016.

[108] Godfrey Boyle, 'Solar Photovoltaics' in Godfrey Boyle (ed) *Renewable Energy: Power for a Sustainable Future* (2nd ed, 2004) 66, 91.

come a near perfect environment for solar energy generation. In addition to high levels of insolar radiation, it experiences the cooling breezes of the coastal trade winds, which keep its atmospheric temperature much lower than generally found at the 25° south latitude.

Figure 16 shows Carnarvon's annual mean maximum daily temperature graphed over a period of 70 years. According to BOM statistics, the mean maximum annual temperature is 27.3°C.[109] Of interest is the general rising trend of annual mean maximum temperature from 25.8°C in 1951 to 28.9°C in 2013. This book does not suggest that the annual mean maximum rise in temperature over the 62 years is 3.1°C, but an average trend rise of around 1°C is a reasonable assertion.

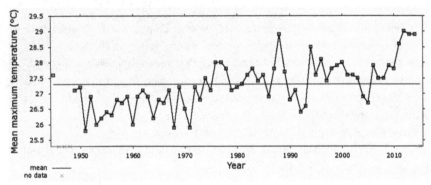

Figure 16: Carnarvon Airport's annual mean maximum temperature: 1945–2015. (Source: Australian Government Bureau of Meteorology and Climate Data Online[110])

To indicate a comparison for the Gascoyne Region generally, Figure 17 shows the annual mean maximum temperature over a similar period for Gascoyne Junction, some 156 km east of Carnarvon, which is 32.1°C.[111]

[109] Australian Government Bureau of Meteorology, above n 92.
[110] Ibid.
[111] Ibid.

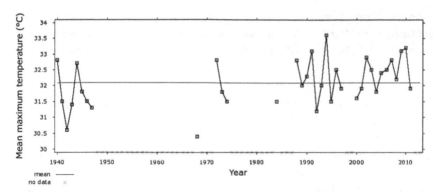

Figure 17: Gascoyne Junction's annual mean maximum temperature: 1940–2015. (Source: Australian Government Bureau of Meteorology and Climate Data Online[112])

Figure 17 also shows a similar trend in the rise of annual mean temperature for the Gascoyne Junction. It is suggested that the higher annual mean maximum is higher than Carnarvon due to its distance of 160 km from the sea and its stabilizing influence on variations in climate.

Despite the data from the Gascoyne Junction being incomplete, the general trend of an increase in temperature is reasonably clear. These trends are generally supported by the estimated trends of rising average surface temperatures suggested in the Brundtland Report "of somewhere between 1.5°C and 4.5°C."[113]

Interestingly, while there appears to be a trend of increasing temperature, the data examined do not appear to indicate a change in average rainfall over the same periods.

Comparative analysis of trends in temperature increases for the broader regions of Western Australia, and the continent in general, is outside the scope of this book. However, BOM statistics indicate that the general trend of increasing temperatures observed for Carnarvon and Gascoyne Junction are also observed for Perth, Western Australia; Darwin, Northern Territory; and Alice Springs in Central Australia; Sydney, New South Wales; Melbourne, Victoria; and Hobart, Tasmania. There is evidence of a lesser increase in temperatures for Adelaide, South Australia; Broome, Western Australia; and Brisbane, Queensland, which do not appear to reflect the trend in increasing temperatures observed by the other locations.[114]

---

[112] Ibid.
[113] Brundtland, above n 15, 148.
[114] Australian Government Bureau of Meteorology, above n 92.

An overarching climatic factor of the climate of Carnarvon, the northwest of West-ern Australia and Northern Australia generally, is the perennial threat of tropical mon-soons or specifically tropical rotating storms known in Australia as *tropical cyclones*.[115] The cyclones bring much wanted rain to the region. However, they also bring devas-tating tempests and flash flooding. The damage can be considerable: lives are lost, stocks destroyed, and communities isolated for days, sometimes weeks, due to dam-age to transport infrastructure.[116]

As indicated in Figure 18, the map of cyclone tracks shown below, tropical cyclones can cross the coast anywhere between Carnarvon and Broome, but the Pilbara region is under the greatest threat. As many as six cyclones a year can occur in the Pilbara. Further south, however, Carnarvon may not be subject to cyclone events for many years.

Selected Tropical Cyclone Tracks

A	Ada	1970	L	Aivu	1989
B	Althea	1971	M	Onson	1989
C	Wanda	1974	N	Nancy	1990
D	Tracy	1974	O	Joy	1990
E	Trixie	1975	P	Fran	1992
F	Joan	1975	Q	Annette	1994
G	Ted	1976	R	Bobby	1995
H	Hai	1978	S	Chloe	1995
I	Hazel	1979	T	Rewa	1994
J	Kathy	1984	U	Ethel	1996
K	Winifred	1986	V	Olivia	1996

Figure 18:   Map of selected cyclone tracks in Australia: 1970–1996. (Source: Emergency Management Australia[117])

Despite the limited occurrence of cyclones in Carnarvon, the solar farm had to con-sider the real threat that it would be subject to at least two cyclone events in its op-erating lifespan, estimated to be at least 20 years.

Historical evidence appears to support that consideration. Despite being estab-lished in 2005, it was not until 10 years later, on March 13, 2015, that the Solex project

---

[115]   Sturman and Tapper, above n 101, 206.
[116]   Some cases include the following: tropical cyclone "Tracy," Darwin, December 1974, 65 killed, 650 in-jured, 35,000 evacuated, $837 million insured damage total estimate over $4.1 billion; tropical cyclone "Bobby," Onslow, February 1995, seven killed; tropical cyclone "Olivia," Pannawonica, April 1996, 10 injured, power installation and 55 houses destroyed; tropical cyclone "George," Fortescue Metals campsite south east of Port Hedland, March 2007 two killed, 28 injured and $8 billion insured damage (Emergency Management Australia, *Hazards, Disasters and Survival: A Booklet for Students and the Community*, (1997) 21–25.
[117]   Ibid, 22.

was first subjected to a cyclone in the form of Tropical Cyclone Olwyn. The Solex project experienced wind gusts in excess of 140 km/h (75 knots) for a number of hours.[118]

For a detailed examination of the formation and factors effecting the courses of cyclones on the Western Australian coast refer to Sturman and Tapper.[119]

3.3   HISTORY OF LOT 42 BOOR STREET CARNARVON

Description of the Land

In 1839 Lieutenant, later Sir George, Grey discovered the mouth of the Gascoyne River for the British. Colonial records and Aboriginal folklore suggest he discovered the Indigenous inhabitants, the Gnulli people, at a camp on the banks of what is now known as "Chinaman's Pool" at the Gascoyne River.

The Gnulli people had occupied the land they refer to as Mungullah for many thousands of years before being "discovered" by Grey in 1839. It is reasonable to note that, as with all Indigenous peoples of the world, they were not "lost" and, therefore, could not be "discovered." Rather, their existence was simply unknown to Europeans, and in particular the British Colonists.

The story of the relationship that developed between the Fitzpatrick family and the Yamatjii people is related in Fitzpatrick's book[120] and outside of the scope of this book, other than to recognize the Indigenous owners of the land on which the Solex project is located—Grey's Plain or Mungullah.

The first European to acquire the land by alienation from the Crown was Robert McAllister on February 20, 1917. Of interest is that his occupation disclosed on the Certificate of Title is that of a farmer.[121] Despite owning the property for some 10 years before his death, McAllister never developed the land or ever farmed it. He remained the registered proprietor until Fullarton claimed it by adverse possession[122] some 75 years later. It was never claimed by his heirs or successors.

The photograph taken of the land in the summer of 1994, shown in Figure 19, presents a fair description of the arid landscape acquired by McAllister in 1917 and subsequently abandoned by him and his successors. A photograph of the same area taken

---

[118]   The Australian Bureau statistics indicate the maximum wind gust at the Carnarvon airport 1 km away was 146 km/h (79 Knots). Australian Government Bureau of Meteorology, *Carnarvon, Western Australia March 2015 Daily Weather Observations* <http://www.bom.gov.au/climate/dwo/IDCJDW6 024.latest.shtml > at March 29, 2015.

[119]   Sturman and Tapper, above n 101, 196–200.

[120]   Fitzpatrick, above n 1.

[121]   Western Australia Certificate of Title Volume 664 Folio 74.

[122]   An occupation or possession of land by someone who has no lawful title to it, which, if unopposed for a certain period, extinguishes the right and title of the true owner. Yallop et al, above n 17, 16.

in 2009 is illustrated in Figure 53, in Chapter 4, for comparison after remedial environmental work had been undertaken. Fullarton, however, saw this land not as "waste country" but rather as "country wasted."

Figure 19: A view across Gascoyne Location 42, January 1994. (Source: Photograph taken by Alexander R Fullarton 1994)

No one had assumed ownership after McAllister's death in 1927. An indication of its value to the wider world was such that, as a result of the failure to pay taxes attributed to the land, a caveat was lodged on the title to secure the debt in 1944.[123]

That caveat remained on the title until the outstanding debt was satisfied by Fullarton, 45 years later in 1989. It appears even the Government of Western Australia was not interested in resuming the land for outstanding liabilities to the Crown. It was on that premise that Fullarton sought to acquire the land at Lot 42 Boor Street Carnarvon, Western Australia.

Acquisition of the Land

Litigation commenced to have the land transferred to Alexander Fullarton as the registered proprietor in 1996. It was initially considered that acts of possession generally by trespass to acquire the land that the Solex project occupies began about 1979.[124]

---

123   Western Australia Certificate of Title Volume 664 Folio 74.
124   Fullarton v The Estate of McAllister (Unreported, Supreme Court of Western Australia, Bredmeyer Master), March 17, 1997, 6.

However, those actions were not considered materially sufficient to establish a solid basis of occupancy by which an order to secure a transfer of ownership by way of adverse possession against a "missing owner"[125] could be obtained.

It was found by the Supreme Court of Western Australia (WASC) that that unequivocal possession commenced sometime in mid-1989.[126] The Court suggested that it was not until around August of 1988 that further acts necessary under the Law to procure the abandoned land—pay land taxes, rates fencing, and the like, to demonstrate to the wilder world that someone was taking possession, had commenced.[127]

Fullarton continued in possession of the land as a squatter[128] in the intervening years until a second claim was lodged with the WASC in 2002.[129] The rates and taxes were paid by the occupant, and all activities carried out to ensure that the registered owner, or his descendants and beneficiaries, could be in no doubt that possession was taking place: fencing and like activities including the erection of a building thereupon.

Action was taken to identify and locate the missing owner. [130] However, no one challenged the activities of the squatter or attempted to evict him and claim their land. That was not really surprising, given the burden of ownership of a landlocked piece of desert. The registered owners were assumed to have abandoned it, and it may have been assumed that their attitude was that if another person wanted it, he could have it.

The intervening period was spent gaining better and more official access, trees were planted and alternative land uses trialled. Those efforts were in vain. The sheep pastured there were killed by marauding dogs, and the trees died—nothing prevailed on that worthless piece of desert.

Part of the findings of the court in 1997 referred to the erection of a house as evidence of possession.[131] To address that finding, a significant action taken to announce a person was in occupation of the land, and, therefore, a challenge to the registered proprietor was the erection of a building thereupon.

Fullarton and his sons, with the assistance of friends, constructed a shed on the land. It was a simple timber-framed shed with corrugated iron cladding as they were

---

[125] Pamela O'Connor, "The Private Taking of Land: Adverse Possession, Encroachment by Buildings and Improvement Under a Mistake" (2006) 33 *University of Western Australia Law Review* 31, 35.
[126] Fullarton v The Estate of McAllister, above n 125, 8.
[127] Ibid.
[128] Someone who occupies a building without right or title. Yallop et al, above n 17, 1184. For further information as to squatters and the adverse possession of land, refer to O'Connor, above n 126, 32.
[129] Fullarton v The Estate of McAllister (Unreported, Supreme Court of Western Australia, Sanderson Master), April 17, 2002, 1.
[130] O'Connor, above n 126, 35.
[131] Fullarton v The Estate of McAllister, above n 125, 8

constructed in the Colonial times but it served the purpose, to give notice that some-one was in possession of the land. As at the end of 2015, it remained intact despite floods, termites, and tempests.

Five years later, in January 2002, a second claim to the Supreme Court was made based on the findings and opinions of the first claim.[132] That claim was successful. On the 17th of April 2002, judgment to acquire title to the land was awarded.

The transfer of proprietorship was not the end to the story of acquisition. In the intervening 85 years from the initial land grant in 1917, Lot 42 had become "land-locked." Access in 1917 was by way of what is now the Northwest Coastal Highway on the eastern boundary.

However, since that time, a number of floods caused by the Gascoyne River over-flowing it banks had resulted in a reasonably significant water way to be developed between the boundary of Lot 42 and the formed roadway constructed at the foot of the range in the early 1960s. That creek had become a barrier to access and an alter-native had to be officially recognized, other than trespass across neighboring land.

Application for access by way of a lane from the northwestern corner of Lot 42 to Boor Street was requested on March 17, 2003. Initially, access was granted by way of a lease of a 12-m strip excised from the neighboring Lot 38, and on November 5, 2009, that portion of land was purchased and amalgamated with Lot 42 to become Lot 100, as illustrated in Figure 20. The process had taken over six and a half years and a con-siderable amount of negotiation with government instrumentalities. The entire pro-cess of acquisition of the land had taken a mere 20 years from inception.

---

[132]   Fullarton v The Estate of McAllister, above n 125.

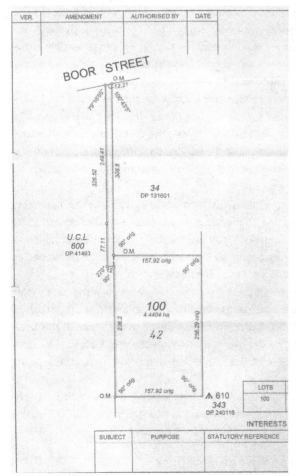

Figure 20:   Surveyed diagram of Lot 100 Boor Street Carnarvon. (Source: Western Australian Record of Certificate of Title Deposit Plan 64248 Certificate of Title Volume 2727 Folio 873)

Using the Land

As a public accountant and registered taxation practitioner, Fullarton knew that the land had to earn income or it would just be another burden without benefit. After a night of heavy drinking to celebrate the "victory," it was considered "oh just great, what am I going to do with a sun-baked piece of dirt that is completely isolated and barren." "What would I ever farm on that?"

The words sun and farm instilled the concept—a solar farm. At that time, the use of solar pv panels as an energy-harvesting system had been established by NASA. Solar panels powered NASA's Space Lab and were scientifically proven as feasible. However, could they be used in a commercial environment to be proven economically feasible in Carnarvon, Western Australia? The investigation began.

Despite not being the registered proprietor but having been awarded the judgement to acquire the title to the land, Fullarton began to gather information and data to construct his venture in 2002. On January 3, 2003, the Certificate of Title Volume 664 Folio 74 was received. Alexander Fullarton had become the registered proprietor of his own piece of worthless desert and the investigation to develop what was to become the Solex project accelerated.

Fullarton sourced data from the published information available at the time to establish a basic business model and an indication of estimated production and fiscal viability. According to Twidell and Weir "the photovoltaic effect [used to generate significant electric power] was discovered by Becquerel in 1839 but not developed as a power source until 1954 by Chaplin, Fuller and Pearson."[133]

The physical characteristics of generating electric current through the use of solar cells is relatively complex and is addressed in detail by Twidell and Weir.[134] Discussion of detailed solar radiation absorption processes is outside of the scope of this book and is largely irrelevant to this study. "Anyway, it is only the physicists who become bothered by such conventions; electrical power engineers just make the systems work!"[135]

This section now looks at the solar insolation,[136] or solar energy harvesting environment, of the location of Carnarvon Western Australia and specifically that of the land containing the Solex project at 118 (Lot 100) Boor Street Carnarvon.

3.4 CARNARVON'S SOLAR ENERGY RESOURCE

For the purposes of this study, the following definitions are made to distinguish between the intensity of solar radiation, or rate at which solar radiation arrives at a surface on the Earth, and the total amount of solar energy received at a particular point over the duration of a solar day.

---

[133] Twidell and Weir, above n 13, 182.
[134] Ibid 182–236.
[135] Ibid 197.
[136] Twidell and Weir define *Insolation* as "the total energy per unit area received in one day from the sun." Ibid, 91.

Iqbal considers the terms *irradiation, insolation, radiation, irradiance, radiance, intensity, radiant flux,* and *radiant flux density.*[137] While he defines radiance, intensity, and radiant flux, he uses the terms irradiation and insolation interchangeably.

Later researchers such as Twidell and Weir make a distinction between *irradiance* or radiant flux expressed as a ratio of Watts per square metre (W/m^2) and *insolation* being "the total energy per unit area received in one day from the sun" (W/m^2/d).[138] The definitions were not known to the developer of the project in 2002; however, for convenience, the distinctions will be used in the following discussion.

Figure 21 is a map of the world showing four insolation radiation zones. It was obtained to give an indication of energy levels that might be expected at Carnarvon in the northwest of Western Australia. Carnarvon is situated on the coast near the western most point of the Australian continent. The map indicates that Carnarvon is in Zone 5, a region that receives around 5–6 hours of productive solar radiation per day.

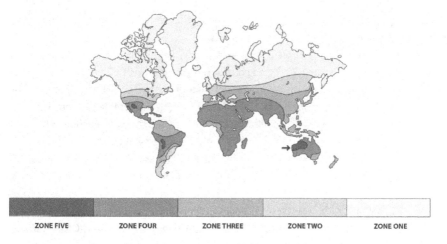

| ZONE FIVE | ZONE FOUR | ZONE THREE | ZONE TWO | ZONE ONE |

Figure 21:   World zones of daily radiation performance. (Data: Peters[139]. Illustration: Moondust Design.)

Of significance to this discussion is that the standard of irradiation used to compare solar panel performance is the power output at an irradiance level of 1000 W/m^2 at an ambient atmospheric temperature of 25°C. System designers calculate variations in expected energy harvest and equipment integration on expected variations from that standard.

---

[137]   Muhammad Iqbal, *An Introduction to Solar Radiation* (1983) 41.
[138]   Twidell and Weir, above n 13, 91.
[139]   Jen Peters *Solar Energy* (2003) University of Wisconson-Eau Claire <http://academic.evergreen.ed u/g/grossmaz/petersj.html> at April 19, 2015.

Unfortunately, actual solar irradiance data for the region was unavailable in 2003. In 2003, there was little known evidence to support the scientific estimations with actual collated field data. However, contact was made with the Birmingham University. The university has a solar observatory located less than 1 km from Lot 42 and gave reasonably reliable data appropriate to projections of solar energy harvest for the Solex Carnarvon Solar Farm.

The university operates a solar research program called the Birmingham Solar-Oscillations Network (BiSON) to provide

> round-the-clock monitoring of the globally coherent, core-penetrating modes of oscillation of the Sun. BiSON consists of a network of six remote solar observatories monitoring low-degree solar oscillation modes. It is operated by the High Resolution Optical Spectroscopy group of the School of Physics and Astronomy at the University of Birmingham, UK. We are funded by the Science and Technology Facilities Council (STFC).[140]

The university's research does not specifically collect data as to irradiance or daily insolation; however, it was suggested by one of the researchers that, if a solar farm was not going to work in Carnarvon, it would not work anywhere.

Schedule 5 of the *Renewable Energy (Electricity) Regulations 2001* (Cth) indicated that Carnarvon (Postcode 6701) was considered to generate 1.622 MWh per annum for every 1 kW of solar panel (photovoltaic) system installed.[141] The production rate of 1.622 MWh per annum for every 1 kW of solar panel (photovoltaic) system installed is deemed to be accurate for the allocation of carbon credits by the Office of Renewable Energy Regulator [ORER; now named the Clean Energy Regulator (CER)].

That legislation indicated that solar pv installations should generate around 4.5 kWh/kW per day. That average is subject to daily weather fluctuations; however, it did give an indication of expected solar energy harvest. In 2003, there was no actual data recorded at Carnarvon to support that estimation.

In 2004, the Western Australian Sustainable Energy Development Office (SEDO) suggested "the average daily solar radiation for Carnarvon is approximately 6.5 kWh/m^2/day (on a tilted [equatorially parallel] surface)."[142] It estimated "[t]he potential output for a 2 MW fixed photovoltaic array would then be around 10MWh per day, after allowing for array and power conversion losses of 25% (sic)."[143] That estimate equates to a generation rate of 1.825 MWh per annum for every 1 kW of solar

---

140   University of Birmingham *BiSON Background: An Overview* (1025) University of Birmingham <http://bison.ph.bham.ac.uk/index.php?page=bison,background> at April 20, 2015.

141   *Renewable Energy (Electricity) Regulations 2001* (Cth) Reg 20(4).

142   Letter from Evan Gray, Senior Program Officer, Western Australian Sustainable Energy Development Office to Alexander Fullarton, April 22, 2004 (held by author).

143   Ibid.

panel (photovoltaic) system installed, some 12.5 per cent higher than that accepted as accurate by ORER. No authority was provided to support that estimate.

Data obtained from the Perth Observatory, in Western Australia, revealed daylight hours at Carnarvon (Latitude 24.880S; Longitude 113.970E) varied from 13h40m on January 1 to 10h35m on June 21.[144] The average is just over 12 hours. The discussion of the variation between a "standard day" and a "solar day" is beyond the scope of this book. Twiddle and Weir discuss that concept for further reference.[145]

Of significance to the harvest of "useful sunlight" is the effect of the angel of solar rays on the surface of a solar panel. Later technological developments in solar panel manufacture have rendered solar panel performance to be less influenced by increasing panel angle. However, in 2003, panel performance was greatly affected by the angle of incidence.

In 2003, the Newcastle Upon Tyne Royal Grammar School conducted an experiment as to the effect of panel angle to direction of light source, the angle of incidence. The panel angle table of findings is shown in Table 3. It indicates that, at that time, panel performance fell significantly after 30° deviation from the right angle to the panel surface.

Angle of deviation (°)	Potential difference (V)	Angle coefficient (%)
0.00	1.50	100.00
5.31	1.45	96.67
10.67	1.44	96.00
16.13	1.43	95.33
21.74	1.43	95.33
27.58	1.42	94.67
33.75	1.40	93.33
40.40	1.37	91.33
47.47	1.31	87.33
56.44	1.18	78.67
67.81	0.92	61.33
90.00	0.51	34.00

Table 3: Panel angle table of findings. (Source: Adapted from Newcastle Upon Tyne Royal Grammar School, *Panel Angle experiment 2003*[146])

Therefore, while locations at Carnarvon's latitude experience an average of 12 hours of daylight on any given day, the angles of the altitude of the sun at sunrise and sunset greatly reduce energy harvest by a fixed panel array. It follows that "useful sunlight" is considerably less than total daylight hours. According to the scientific support for

---

[144] Facsimile from Perth Observatory to AR Fullarton, February 26, 2003.
[145] Twidell and Weir, above n 13, 435.
[146] Newcastle Upon Tyne Royal Grammar School *Panel Angle* <http://www.rgs.newcastle.sch.uk/nsep/So lar%20Energy/angle/angle.html> at July 4, 2004.

the renewable energy legislation, that loss is considered to be as high as 7h30m. That theoretic expectation is compared to actual data presented in Chapter 4.

One of the climatic considerations for calculation estimated solar pv energy harvest is ambient air temperature. Data provided by the BOM indicates that Carnarvon has a relatively stable temperature.[147] Its maximum temperatures range from an average of 32.6°C in February to an average of 22.1°C in July with an annual average maximum temperature of 27.1°C.

It is interesting to note that that temperature range is very close to standard solar panel testing conditions of 25°C, as discussed earlier. The data also show that there is no record of hail, 212 clear days, only 50 cloudy (or overcast) days, and the balance of cloud cover average is just over a quarter of the sky. Climatically, therefore, Carnarvon is a near perfect location to harvest solar energy using pv solar panels.

However, there was considerable resistance to the concept of a large grid-connected solar farm by the Western Australian State-owned electricity utility Western Power. To establish a reference point for discussion of the capacity of the intended solar farm, Fullarton simply divided the area of Lot 42 by an estimate of area required by a solar chassis. An estimate of 2 MW was established.

That proposal met with amusement by representatives of Western Power. In 2003, Western Power representatives suggested that, while the project was technically feasible, and that up to 6.3 Gigawatt Hours (GWh) of solar energy could be taken up by the Carnarvon Town distribution system annually, a capacity estimated to be twice that of the proposed 2 MW solar farm, it was evaluated that the project was not economically feasible.[148] Therefore, such a considerably sized solar farm could never be considered. That evaluation is examined in detail in Chapter 4, where it is compared with the actual fiscal outcomes of the project.

In 2004, SEDO detailed its evaluation based on the construction of solar pv installations at Singleton and Queanbeyan in New South Wales. That evaluation foresaw many matters that were required to be addressed. It also implied the project would not be economically feasible and went further to suggest the Carnarvon town distribution system could not accommodate a 2-MW PV array.

It stated:

Given the average electricity demand in Carnarvon is around 5MW, there may be time when the potential output of a 2 MW PV array would be too great and some potential output would need to be "dumped." The output of the PV system that may actually be used is therefore

---

147 Facsimile from Bureau of Meteorology to AR Fullarton, February 21, 2003.
148 Letter from Kim Varvell, Commercial Manager Regional Branch Western Power to Alexander Fullarton, May 8, 2003 (held by author).

likely to average around 8–9 MWh per day. The annual output would then be around 2.9–3.3 GWh.[149]

The document does not provide an authority or any other support for that evaluation. It is also noted that SEDO's estimation of solar energy that could be taken up by the Carnarvon Town distribution system was around half that estimated by the utility itself. Perhaps, SEDO was simply being conservative in its assessment. SEDO's evaluation will be examined in detail later in this chapter under the heading "Financial Consideration."

In order to provide a basis for evaluation of actual energy harvest and to provide an indication of economic viability, it was agreed by engineers at Western Power that a 1.92-kW grid tied installation at 3 Crossland Street in December 2003 be established as a trial plant to obtain actual data of solar harvest. It was then agreed that a 15-kW system could be trialled to provide further data as to the impact of solar pv on Western Power's Carnarvon Town distribution system.

The proposal for the construction of that initial 15 kW system is contained in Appendix A. It will be used as a basis for further discussion in the following chapter for comparative analysis with actual production of the Solex project.

3.5   SOLAR FARM CONSTRUCTION

A key influencing factor in the decision to construct a solar farm in Carnarvon was the knowledge that the arid landscape and desert-like climate resulted in predominantly clear skies and high solar radiation levels. Fullarton's knowledge of the climate from his long-term residency and family background supported the belief that Carnarvon might be an ideal location for a solar farm.

Although folklore and local experience may support a qualitative assertion, it cannot substitute quantitative scientific research for empirical analysis. However reliable data and practical support as to design of solar pv systems was almost nonexistent in Western Australia in 2002. The concept of generating electricity using pv solar panels was known to the scientific, space, and academic communities, but the general Australian community was largely ignorant of harvesting solar energy through the use of solar pv systems.

To illustrate the lack of industry and community knowledge of solar pv systems in Western Australia, at that time, the following analysis of renewable energy generation data from the Australian Government's Clean Energy Regulator's Office Renewable Energy Certificate Register is presented.

---

[149]   Letter from Evan Gray, Senior Program Officer, Western Australian Sustainable Energy Development Office to Alexander Fullarton, April 22, 2004 (held by author).

Despite the introduction of the *Renewable Energy Electricity Act* (Cth) in 2000, by 2002, only four solar pv systems had been installed in Western Australia[150] and 20 throughout Australia[151] including installations at Singleton and Queanbeyan in New South Wales, which were referred to by SEDO in its evaluation of the Solex project in 2004, both of which were installed prior to 2000.

The Renewable Energy Certificate Registry records show that five only large-scale generation certificates (LGCs), representing 5 MWh, had been created for the 2001, 2002 years throughout Western Australia and 1,386 MWh for the entire nation.[152] At that time, solar energy was making a very small contribution to Australia's energy-generation capacity.

By the end of 2005, the number of LGCs registered had risen to 25 (equating to 25 MWh) for Western Australia of which the Solex project had contributed 14 (14 MWh) to Australia's energy demand of a total of 5,113 MWh of solar pv energy generated across the nation. In 2005, Western Australia's contribution to Australia's solar-fueled energy-generation energy was less than 0.5 per cent of the Nation's total solar pv capacity, and Solex was contributing over 50 per cent of that Western Australia's contribution.

An attempt to introduce solar pv was made by the Western Australian government earlier, in 1995. Verve Energy, the Western Australian state electricity generation utility, began operating a 21-kW solar pv power system in Kalbarri. At that time, it was the largest solar pv system of its kind to be connected to a major electricity grid in Australia. The system is connected to the main electricity grid in Western Australia, and depending on local weather conditions, it can deliver about 14 kW of electricity or 67 per cent of its theoretic generational capacity.[153] By comparison, the Carnarvon Solar Farm delivers about 37 kW from its 46 kW generational capacity or 80 per cent of its theoretic generational capacity.[154]

Solar Farm Design

A key consideration in the design of the solar farm was the mounting of the solar panels and selection of frame/chassis design. Not only is the position on the Earth in

---

[150] The 21-kW installation at Kalbarri is not included in the Clean Energy Regulator's registry as its installation is prior to 2000, and it is not registered for the purposes of registering RECs.

[151] Some installations are not included in the Clean Energy Regulator's registry as its installation is prior to 2000, and it is not registered for the purposes of registering RECs.

[152] Australian Government, Clean Energy Regulator, *REC Registry* (2014) <https://www.rec-registry.gov.au/rec-registry/app/public/lgc-register> at March 29, 2015.

[153] Western Power, *Solar Power In Kalbarri* (2007) <http://www.worldofenergy.com.au/factsheet_solar/07_fact_solar_kalbarri.html> at June 24, 2010.

[154] Data sourced from Solex Carnarvon Solar farm inverter data loggers.

relation to the Sun (latitude) a critical factor in designing solar farms in Australia[155] but also the constant variations in altitude angle of the Sun caused by seasonal variations, as shown in Figure 22.

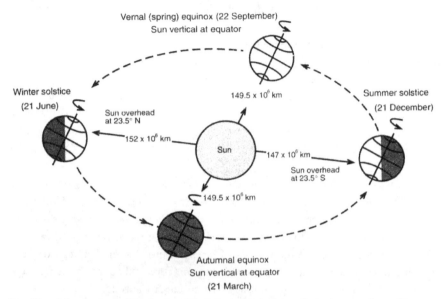

Vernal (spring) equinox (22 September)
Sun vertical at equator

Winter solstice
(21 June)

Sun overhead
at 23.5° N

152 x 10⁶ km

149.5 x 10⁶ km

Summer solstice
(21 December)

Sun

147 x 10⁶ km

Sun overhead
at 23.5° S

149.5 x 10⁶ km

Autumnal equinox
Sun vertical at equator
(21 March)

Figure 22:   Daily and seasonal variations in solar radiation in the southern hemisphere. (Source: Sturman and Tapper[156])

Given the total annual variation is 47° and Carnarvon's latitude is roughly 25° S and, for three months of the year, the Sun rises and sets nearly 22° south of east and west at the summer solstice, it was considered that the solar chassis must have the ability to follow the seasonal variation illustrated. In addition, the perennial threat of tropical cyclones in the region,[157] as illustrated previously in Figure 18, also had to be considered.

The solar pv installation at Kalbarri was inspected to assess the effectiveness of alternative solar panel array systems. It was observed that the installation used a "staggered" system of daily solar tracking frames. That arrangement was possible used as, at the time, it was generally considered by the scientific community that solar panels were particularly sensitive to variations of angle of the Sun's rays. The ability

---

[155]   For a detailed explanation of how latitude influences solar panel tilt, refer to Twidell and Weir, above n 13, 89–95.
[156]   Ibid, 31.
[157]   Tropical rotating storms are generally called hurricanes in the Atlantic, typhoons in the North Pacific, and cyclones in the South Pacific and Indian Ocean. Sturman and Tapper, above n 101, 439.

of follow the sun's path on a daily basis was, therefore, considered essential to high production rates.

However, it was also noted the technology proved unreliable; the trackers often were left facing the setting sun and, hence, failed to reset to face the dawn sun the following morning. The production rates of the Kalbarri installation may have disappointed its developers, but they did assist in formulating a structure of arrays, which later proved to be very successful—Fullarton decided to abandon daily sun tracking systems as an alternative structure.

Investigations of alternative framing structures conducted in 2003 revealed that the Australian telecommunications corporation, Telstra, had been using solar energy in standalone systems to power remote transmission stations around Australia for a number of years. The solar energy was used solely to charge lead acid batteries to power its telephone retransmission installations.

The remote structures were designed to be erected in the Bush using simple tools and to be readily adjustable to various latitudes from Victoria, in the 30°S range, to the far north of the Northern Territory at just over 10°S. Figure 23 shows a side view of the array chassis. Note the multiholed adjusting bar at the lower front to vary the altitude angel according to the latitude of the structure. The modular construction permitted fabrication on an industrial scale with erection and equatorial panel angle to be set in accordance with the specific site according to latitude.

Figure 23:   Telstra solar panel chassis. (Source: Photograph taken by Alexander R Fullarton, 2004)

Of interest was the durability of the Telstra chassis, in particular its ability to withstand cyclonic winds. Telstra staff reported that a set of Telstra solar panels is located at Exmouth Western Australia, which endures very frequent and high-intensity storms. By 2003, the structure had endured no less than six cyclones. One of which, Tropical

Cyclone Vance, a category 5 tropical cyclone, had devastated the town of Exmouth. It was reported that, despite being extremely exposed to the winds, the solar panel structure had suffered no damage.

It was decided to copy the basic Telstra chassis. However, rather than fixing the panel angle to suit the latitude, it was decided that the chassis be adjustable to accommodate the seasonal variations from 45° north-facing at the Winter solstice on June 21 as shown in Figure 21, to 12.5° south-facing at the Summer solstice on December 21.

As shown in Figure 24, the solar panel chassis used by the Solex Carnarvon Solar Farm, the design allows the panel angle to be altered from an angle of 45° north-facing to an angle of 12.5° south-facing. It also provides for the platform to be secured in a horizontal position as an anticyclone measure.

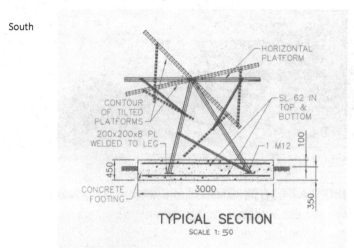

Figure 24:   Solex solar panel chassis. (Source: Stein Halvorsen, Engineered diagram, "Solex Carnarvon Solar Farm: Solar panel chassis", 2009[158])

It is indicated that the altitude of the noonday sun on December 21 at Carnarvon is 88.5°[159]. Theoretically, a flat surface would be preferable. However, field trials later found that the dust build up on a flat surface could not be successfully countered. It was found that a minimum panel angle of 12.5° is required to combat a build-up of dust on the panels that reduces their effectiveness.

---

[158]   Held by author.
[159]   The Solex Carnarvon Solar Farm is located at 24° 54 min S. On December 21, the sun is perpendicular to the Tropic of Capricorn at 23° 30 min S, a difference of 1° 24 min.

As sunrise and sunset occur south of east and west, a south-facing panel angle has a superior harvest during the four-month period from the mid-October to mid-February. Computer modeling using Homer software supports the Solex recorded data as to the higher energy harvest of south facing solar panel structures during those months.[160] Therefore, as illustrated in Figures 25 and 26, the Solex solar chassis are adjustable to allow for seasonal variations but are not able to follow the sun on a daily basis.

Figure 25: Solex panel arrays in mid-winter facing 45°N. (Source: Photograph taken by Alexander R Fullarton, 2008)

Solex data indicate that, by making seasonal adjustments of solar panel angle, the annual energy harvest increases from 1.622 MWh of electricity per kilowatt of solar panels installed[161] to 2.029 MWh per kW of solar panel power. That is, an increase of some 25 per cent increase in energy harvest over a fixed array installation.

The seasonal tracking system provides an increase in energy harvest approximately commensurate with a daily tracking system but removes the complexity of the daily tracking components and their related propensity for mechanical faults.

---

[160] Hybrid Optimisation of Multiple Energy Resources (HOMER) <http://www.homerenergy.com/software.html> at April 27, 2015.

[161] Solex data show a harvest rate of 2.029 MWh/kWpa. The expected harvest considered by the science supporting the *Renewable Energy (Electricity) Regulations 2001* (Cth) Reg 20(4) for the Carnarvon region to be 1.622MWh/kWpa.

Figure 26:   Solex panel arrays in midsummer facing 12.5°S. (Source: Photograph taken by Alexander R Fullarton, 2010)

Figure 27:   The southern azimuth indicated by the setting sun in midsummer. (Source: Photograph taken by Alexander R Fullarton, 2015)

Figure 27 indicates the southern azimuth[162] from west of the sun at sunset. The wind turbines are aligned precisely west to east, which is 270°–90° of true north. The glare

---

[162]   For convenience, the definition of *azimuth* used here is that applied to surveying "an angle measured clockwise from the north" as the Solex project is aligned True North. For navigation purposes, Carnarvon has a magnetic deviation of zero, ie, at Carnarvon Western Australia True North is precisely the same as Magnetic North. For alignment of solar panels, the astronomical definition of *azimuth* should be used

of the setting sun is well to the south of west. By reference to Figure 21 and Carnarvon's latitude of 24° 54 min south, that azimuth is calculated to be 22° 06 min south of west or 247° 54 min of true north.

If the seasonal adjustment is not made, the rising and setting sun would be exposed to the rear of the solar panels for a number of hours per day. The seasonal adjustment allows for that range of azimuth and increases total energy harvest accordingly.

In early 2003, a search of the Internet located a solar panel and equipment supplier in Arizona in the United States of America. Contact was made with that company and the feasibility of importing 2,000 kW of panels and enabling equipment was investigated. In addition, negotiations to connect the solar farm to the Carnarvon town distribution system began with the state-owned electricity utility—Western Power.

The evaluation conducted by Western Power was less than enthusiastic. The full economic cash-flow analysis is contained in Appendix B. The evaluation considers that, while the energy generated by the 2 MW (2,000 kW) solar farm could be incorporated into the distribution system, the project to be economically unviable.

The projected harvest of the 2,000 kW of installed capacity was expected to be

$$2,000 \text{ kW} \times 1,622 \text{ kWh/pa} = 3,244,000 \text{ kWh/pa or } 3.244 \text{ GWh}$$

In its 2003 communiqué, Western Power states that "[a]lthough Western Power will take up to 6.3 GWh per annum, it is more likely the panels will generate the figures shown in the table"[163] (attached in Appendix B).

It is of interest to note that, in 2003, Western Power had little, or no concern, as to the volume of dispersed embedded solar pv, which could be incorporated into the Carnarvon town electricity distribution system. However, it did consider that the economic feasibility of a 2-MW system would be severely impacted by the capital costs of enabling equipment required to disperse a high volume of electricity generated by such a large solar pv system through the high voltage network.[164]

It was also pointed out that

Western Power is obliged to carry spinning reserve or standby generation to cover renewable energy generation. This is to ensure the reliability of the electricity supply to our customers. Hence the solar farm proposal will not defer or displace Western Power capital investment in generating plant and the project merit will be determined by the marginal operating cost of generating electricity.[165]

---

as the sun is a celestial body "commonly reckoned from the south point of the horizon towards the west point." As the Solex project has been laid out by way of survey, reference to True North has been used for consistency and to avoid confusion for the reader.

[163] Varvell, above n 149.

[164] Letter from Grant Stacy, Asset Manager Regional Branch Western Power to Alexander Fullarton, February 18, 2003 (held by author).

[165] Ibid.

At that point, it appears the philosophy of utility administrators was that renewable energy systems were not sufficiently reliable to produce electricity of the standard required to meet the utility's required electricity supply standards. Therefore, the economic savings gained from renewable energy sources, and hence its economic value, was considered to be insignificant.

In Western Power's letter of May 8, 2003, it was stated the approximate unit value to Western Power was around 8.5 c/kWh.[166] No justification was given to support the offered price. However, it is reasonable to assume the reasons outlined in SEDOs letter would have been of considerable influence.

It was stated that

> Residential electricity prices in Western Australia are currently set by the Western Australian Government rather than through competition in the retail market or by an independent regulator. The Western Australian Government subsidises residential electricity prices such that 2013/14 residential prices would need to increase by 30 per cent to reflect the total estimated cost of supply.[167]

It was also suggested that

> Before 2009, prices for residential customers had not increased since 1997/98 (excluding GST). Prices for small business customers had not increased since 1991/92 (excluding GST). Large business tariffs increased in 2007/08 for the first time since 1992/93 (excluding GST).[168]

Therefore, any attempt to quantifiably support the cost of generation, and hence the true economic value of the displaced fossil fuel-generated electricity, would have been a matter of philosophy rather than a justifiable accounting function.

Fullarton assumed that the Western Australian Government could not afford to meet the growing subsidy indefinitely and would have to increase tariffs at some point in the future. That increase ultimately occurred some four years later in 2009.

In 2015, debate as to the true economic value of solar pv-generated electricity despatched to the grid continues; however, it is suggested that the offered feed-in tariff (FiT) paid by Western Australia's energy utilities is far short of its economic value, and certainly extremely poor should the environmental and social costs be considered.

In 2015, Horizon Power[169] reduced its FiT despite increasing its tariffs to consumers for electricity to compensate for rising generational costs. It reduced its purchasing tariff from its comparable domestic tariff rate in 2009, to around 1 cent less than its

---

[166]   Varvell, above n 149.

[167]   Australian Energy Market Commission, *2014 Residential Electricity Price Trends Report* (2014) ix.

[168]   Western Australian Government, Department of Finance, *Electricity Pricing* (2015) <http://www.financ e.wa.gov.au/cms/content.aspx?id=15096> at September 29, 2015.

[169]   In 2006, Western Power Corporation was disaggregated into four separate entities with distinct roles and responsibilities: Verve Energy, generation; Western Power, transmission; Synergy, distribution and sales; and Horizon Power, generation, transmission, and distribution and sale of electricity in regional areas outside that of the South West Integrated System (SWIS), which is in Western Power's jurisdiction.

domestic tariff rate for its Carnarvon customers in 2012, to 10.56 cents in 2015. That reduction in its "buyback" rates was despite the domestic tariff rising to 25.70 cents at the same time.

The ratio between purchase and sale prices of electricity had altered from 1:1, in 2009 to 2.43:1 six years later. Henceforth, the utility was making a profit from the purchase of solar pv-generated energy, as well as benefitting from the cost savings of reductions in fossil-fueled–generated energy that is displaced by the solar pv systems.

Ultimately, prices for Solex's renewable energy were established by negotiation until 2015, when government attitudes toward the economic benefits of renewable energy returned to pre-1997 philosophies toward climate change and economic focus changed.

Resistance to changing technology is not unique to society. An interesting comparative to the mistrust of the reliability of renewable energy sources might be the development and introduction of the steamship. Figure 28 shows the steamship Oceanic under sail with smoke billowing from its coal-fired engine room in the 1870s.

Figure 28:   The Screw Steamship "Oceanic," 1870. (Source: Ramsay Muir, *Bygone Liverpool* (1913)[170])

At the advent of the introduction of steam engines into ships, naval architects continued to construct masts and rigging for sails in the belief that the steam engines could not provide the reliability of wind power required by commercial shipping services.

---

[170]   Ramsay Muir, *Bygone Liverpool illustrated by ninety-seven plates reproduced from original paintings, drawings, manuscripts, and prints with historical descriptions by Henry S. and Harold E. Young* (1913).

In the intervening 100 years of mechanical power, the philosophy had changed from a suspicion of the reliability of mechanical (fossil-fueled) power to a suspicion of the reliability of natural (renewable) energy power sources.

That is particularly interesting, given that Western Power was to later claim grid instability problems, caused by the level of solar pv-generated electricity within the distribution system and ceased to consider further applications for connection to the grid. That is, despite that, by the end of 2010, less than 1 MW of solar pv capacity had been installed, and there had been no changes in capacity of the fossil-fueled power station in Carnarvon.

Therefore, the 2010 limitation on solar pv installations was placed on the Carnarvon town distribution system was at a level of around one-quarter of the generation capacity that the 2003 evaluation indicated that could be accommodated. This estimate is based on the capacity to annual generation rate of 1 kW capacity results in 1.622 MWh of annual generation. Therefore, the estimated annual production of 6.3 GWh of electricity generation would require 6,300/1.622 of solar capacity or 3.884 MW.

Despite the capacity of 2 MW being evaluated as practically feasible, Western Power engineers were sceptical that the fossil-fueled power system could accommodate a renewable energy system of that size. That opinion was supported by the Western Australian Sustainable Energy Development Office, which was more conservative than Western Power.

Negotiations continued for many months until, by July 2004, it was agreed that a "trial plant" of just 15 kW be installed in able to quantify productions rates and disturbance to the distribution system established. The "proposal to connect a 15 kW solar photovoltaic array to Western Power Corporation's Carnarvon low voltage network" at a fixed purchase price somewhat higher than the "standard rate" at the time was confirmed in writing by October 2004.[171] Western Power further stated that it supported "initiatives of this type to increase its renewable energy program [despite it not being] economic."[172]

It was on the premise that a 15-kW system was permitted; of that, a program of works and a schedule of materials could be prepared for costing and finance purposes. This section now considers the financial considerations of the development of the Solex project.

The development of the entire Solex project was conducted on a "trial-and-error" basis in a series of smaller projects over the 10-year period. Financial constraints were

[171] Letter from Mike Laughton-Smith, Manager Regional Branch Western Power to Alexander Fullarton, 21 October 2004, (held by author).
[172] Ibid.

not the only limiting factors. Engineering considerations as well as the impact of possible variations to the Carnarvon Town electricity distribution system also had to be evaluated. Therefore, the development was conducted in the series of stages listed further.

Stage 1   The construction of the 15.8-kW initial solar pv array, 2005;

Stage 2   The construction of a 30.2-kW solar pv array extension, 2007;

Stage 3   The integration of a 5-kW Westwind wind generator, 2008;

Stage 4   The construction and integration of an ice-works, 2009;

Stage 5   The integration of a further two 5-kW wind generators, 2010;

Stage 6   The construction of a farmhouse, 2013;

Stage 7   The replacement of GE 110 W solar panels with Astronergy 270W;

Stage 8   The removal of the wind generators, 2015.

The following evaluation is conducted in order of that list of stages.

Financial Considerations

Stage 1—15.8-kW Solar Array

The Western Power and SEDO evaluations provided a basis for the construction of the 15-kW "trial plant" in 2005. In April 2004, SEDO advised "The RRPGP provides rebates of up to 50% of the cost of renewable energy systems replacing diesel generation [to] 'off-grid' [electricity distribution systems]."[173] It provided an 11-point assessment guideline, which was based on and clarified a general 13-point eligibility guideline, by which it could potentially provide financial assistance for the project under the Australian Government's RRPGP.

By April 2004, a number of considerations raised by SEDO had already been addressed by way of negotiations with Western Power, which had commenced the previous year. It was noted that, despite both Western Power and SEDO being forwarded a copy of the proposal for the 2-MW solar farm concurrently, Western Power's evaluation of the proposed project preceded that of SEDO by some 11 months. Fullarton found that one of the key factors influencing the proliferation of tax avoidance schemes in Australia in the 1970s–1990s was administrative delays by authorities in attending to matters before them.[174] The administrative delays experienced in the

---

[173]   An "Off-grid" electricity distribution system is defined as one being: Isolated or islanded systems (including private systems) or grids that are not connected to the National Electricity Market or the South West Interconnected System [in Western Australia]. AECOM Australia Pty Ltd, "Australia's Off-Grid Clean Energy Market Research Paper" (Research Paper, Australian Renewable Energy Agency, 2014) 1.

[174]   Alexander Robert Fullarton, *Heat, Dust and Taxes* (2015) 154.

processing of matters addressed by SEDO were considerable and, to later, have severe impacts on cash-flows in financial considerations of the project.

In fairness, the application of RRPGP funds to a project of this type was novel, as that program was intended to displace the use of diesel in microelectricity generation systems used by farms and station homesteads in Australia's Outback, not larger town distribution systems such as Carnarvon. It was argued that the principle was the same, that is, the solar farm displaced diesel from a single diesel-powered generation source and that the town of Carnarvon could be considered as no more than a very large station homestead and outbuildings with multiple residences. It took some time for that argument to be settled.

Finance for the project was to be by way of 50 per cent through the RRPGP, accompanied by a 5 per cent contribution by the Western Australian Government[175] and a first mortgage secured loan of 50 per cent from the Bank of Western Australia. As an anecdote of humorous incidents during the project, Fullarton was asked "How are you going to fund this?" The reply was "Well the Federal Government will stump up 50 per cent, the State government 5 per cent and Bankwest will finance up to half of the cost of these sorts of ventures." To which the enquirer replied "And what will you do for the rest of the money?"

As detailed in Appendix A the estimated construction cost was placed at roughly $175,000. A source of solar panels and enabling equipment was found in Western Australia from a company located in Perth.

In addition to the solar panels, fencing, framing, and 15 kW of enabling equipment (grid-tie inverters) as well as metering and feeder connections were required. The land also required some earthwork and specialized trades people to complete the installation. Compliance with regulator bodies for the requisite planning approvals and building licences incurred costs and fees, which were included in that estimate.

The following is the proposed construction budget for the construction of 15 kW of a solar installation in 2004 as listed in Appendix A.

Panels 100@ $850	$85,000
Freight, Insurance	6,000
Transformer/Inverter/Controller	20,000
Metering and connection	10,000
Land clearing and earthworks	2,000
Fencing 300m @ $30m	9,000
Framing	12,500
Incidentals, Engineering, Erection Labor	10,000
Project Management	20,000
Total estimated Construction Cost	**$174,500**

---

[175] In accordance with the RRPGP—Business Remote Area Power Supply Program requirements as at July 2004.

The RRPGP prepurchase application form was slightly different from that at $172,500 as solar panels had become cheaper in the interim. It is also noted that the expected solar array output was estimated to be 65 kWh/day. The actual energy harvest is discussed in the following chapter, which outlines the actual fiscal outcomes of the project.

Complication arose as to the source of solar panels and enabling equipment in that the originally approved supplier ceased business during the approval and construction process. An alternative supplier had to be sourced with different makes and models of solar panels and inverters.

The final cost of construction subsequently increased to $199,339 and the final capacity to 15.84 kW ($12.58/W). Of the cost of $199,339, the RRPGP rebate contributed $104,007 or 52 per cent and Fullarton the remainder by way of personal funds, work-in-kind, and borrowings.

Stage 2—30.2-kW Solar PV Array Extension, 2007

The 15.8-kW solar generation system proved to be successful. Western Power had noted no disturbances to its network, and in November 2006, it was decided by both Western Power engineers and Solex to expand the installation by a further 30 kW.

The same RRPGP and private borrowings financial resources were applied. It was estimated that the extension would cost $319,400 of which SEDO committed $175,120 of the RRPGP rebate system through the Business RAPS program as before. Processing of the application had taken just 34 days. After consideration time taken away from administrators due to the Christmas and New Year Holidays, that clearly indicated the administrative system had benefited from the "pioneering" application of the two years' previous.

The extension was commissioned on August 8, 2007, at a total cost of $281,071 ($9.31/w) of which the RRPGP component was $153,393.

Stage 3—5,kW Westwind Wind Generator, 2008

In November 2007, after examination of solar performance and energy harvest by engineers of Horizon Power (formerly Western Power), it was decided that a trial could be conducted to integrate wind power generation into the solar pv, enabling equipment to extend to duration of electrical generation into the hours of darkness. It was also intended to ascertain Carnarvon's wind resource beyond that of scientific estimates.

A 5-kW Westwind wind generator was selected as that capacity could readily interface with the existing Fronius 5-kW solar inverters. The estimated cost was $41,021 to an additional 85 kWh per day or 31 MWh per annum. Some modifications were required to enable the mast to be totally lowered as an anti-cyclone measure that resulted in the completed cost of $46,372 with an RRPGP contribution of $23,186.

The wind generator was commissioned on August 12, 2008. The entire process from conception to commission had taken just 9 months compared to 2 years for the initial 15-kW stage in 2005.

Stage 4—The Construction and Integration of an Ice-Works, 2009

In January 2009, it was decided to construct and ice-works to "value-add" to the electricity generated by the solar farm. The commissioned cost of that operation including plant vehicles, freezer, and buildings cost $324,387 financed by borrowings from external sources.

The ice-works consumes around 100 kWh per day or that energy harvested from a 15-kW of solar pv array. In addition to the solar array at the solar farm, the ice-works building has an integrated 4.6-kW roof mounted solar pv array to supplement its draw from the solar farm. There were no rebates or subsidies for this section of development.

Stage 5—Two 5-kW Wind Generators, 2010

In order to fully exploit the remaining two 5-kW Fronius inverters, two further Westwind 5-kW wind generators were added in 2010. The constructed cost was estimated $108,482 with an RRPGP component of $54,241. The estimated energy harvest was less ambitious, as there were actual data to support anticipated energy production that had been derived from the production rates of the initial 5-kW wind generator. It was anticipated that energy production was likely to be 17 kWh per day in place of the original 85 kWh of the first machine.

The final constructed cost was $100,000 with an RRPGP component of $50,000.

By 2010, the RRPGP rebate program had contributed $330,586 of the $626,782 capital cost of 46 kW of solar pv array and 15 kW of wind generator capacity. Without that financial support of the Federal Government of Australia, the Solex project would not have been possible.

Stage 6—The Construction of a Farmhouse, 2013

To reduce fossil fuel consumption by way of travel to and from the Solex project, it was decided to relocate to the farm. A farmhouse was constructed, which also incorporates a 5.5-kW roof-mounted solar array at a cost of $311,500. The cost was partially funded by borrowings of $215,000. There were no rebates or subsidies for this section of development.

Stage 7—The Replacement of GE 110-W Solar Panels with Astronergy 270W Panels

The 15.8-kW array was increased to 21.2-kW in September 2014 by exchanging 128 GE 110-W solar panels of the original array with 72 Astronergy 270W at a cost of $13,860 ($0.71/W). There was no recovery value from the technologically obsolete GE 110-W panels.

Stage 8—The Replacement of the Wind Generators, 2015

In March 2015, the wind generators were lowered in preparation for Tropical Cyclone Olwyn. The data collected over five years shows that the generation of the wind generators was significantly below the scientifically based harvest estimates. Collectively, the wind generators harvested just over 50,000 kWh in five years, or an average of 27 kWh/day.

The Solex project data indicate that a similar-sized solar pv array harvests an average around 87 kWh/day. The wind generators will be replaced by a 15-kW solar, which will integrate with the existing circuitry and enabling equipment. It is estimated that the construction cost will be less than $20,000. There may also be some recoverable value in the dismantled wind generators.

Overall, as on September 2015, the Solex project consisted 61.46 kW[176] of solar pv arrays, 15 kW of wind generation, an ice-works, and a farmhouse cost $1,276,529 plus the acquisition costs of the land and the purchase of the access lane. Federal government funding by way of the RRPGP had contributed a total of $330,586.

Table 4 details the progressive costs of each stage of the Solex renewable energy generation and shows the total costs as well as the average cost per watt of the installations. As to the solar pv capacity, the 61.46 kW had cost a total of $512,475 or $8.34/W. Federal subsidies reduced that cost to $4.19/W.

Stage	Constructed cost	Rebate	Net cost	Capacity (kW)	Net $/watt (after rebate)
1	$199,339	$104,007	$95,332	15.8	6.034
2	$281,071	$153,393	$127,678	30.2	4.228
3 (wind)	$46,372	$23,186	$23,186	5.0	4.637
4	$9,800	$0	$9,800	4.5	2.177
5 (wind)	$100,000	$50,000	$50,000	10.0	5.000
6	$8,405	$0	$8,405	5.5	1.528
7(1a)	$13,860	$0	$13,860	5.4	2.567
8[177]	$20,000	$0	$20,000	15.0	1.334
Total	$678,847	$330,586	$348,261	76.4	4.499

Table 4:   Solex renewable energy generation constructed costs: 2005–2015.

---

[176]   This capacity excludes 15 kW of wind generation to be replaced by a solar pv array of approximately the same capacity that is yet to be installed.

[177]   Under construction to replace wind generation; therefore, this does not increase overall capacity.

That remains expensive compared to the cost of solar installations in modern times. In addition, countless hours of unpaid work were contributed by Fullarton and his family. However, that only substantiates to old maxim—"There is no money in YOUR sweat." Such is the intrinsic value of pioneering and the privilege of the bold claim of being "first."

In October 2004, an agreement was reached with Western Power to construct a

> 15 kW renewable energy generation system at Lot 42 Boor Street Carnarvon, in Western Australia at a fixed purchase rate of 12c/kWh of energy purchased and [Western Power] reserve[d] the right to purchase the Renewable Energy Certificates ("RECs") accruing from the 15 kW renewable energy generating system at prices to be negotiated at the time the RECs [were] realised. At this price, WPC will purchase a maximum of twenty-five thousand kilowatt-hours (25,000 kWh) per year of contract from the 15 kW renewable energy generating system.[178]

The estimated cash flow forecast for a year had evolved to the following

Annual generation

15 kW × 1.622 MWh/kW = 24,330 kWh @ 12 c/kWh =		$2,920
	24 RECs @ $35 =	$840
Total revenue		**$1,680**

less Operating expenses

Interest $78,500 @ 10 %	$7,850
Rates and taxes	$900
Total expenses	**$8,750**
Projected net cash loss	**$7,070**

It appeared the estimate of Western Power was going to be correct. However, at least the solar farm would be revenue producing and create implications of a tax benefit. It is pointed out that the prime directive of the 15-kW generation system was not to return an economic benefit but rather to trial an integrated grid-tie system to establish the reliability of solar pv. The primary function was to address the concerns of the utilities that the integration of solar may, in some way, have a negative effect on their energy generation and distribution system. The actual production and fiscal outcome will be discussed in the following Chapter 4.

---

[178] Laughton-Smith, above n 176.

By May 2005, Western Power had relented from its firm stance on the integration of renewable energy to its distribution network and the final agreement permitted the maximum capacity to be increased to 105 kW.

However, by the end of November 2004, pre-purchase application approvals had been complied with from SEDO, the local authority, and other relevant authorities; the purchase of equipment and construction could commence. Site works were carried out on December 9, 2004—the Solex project had begun.

Construction

Stage 1—15.8-kW Solar Array

The system design to integrate the solar panels with the enabling equipment sourced in Western Australia was certified by Mr Durmus Yildiz of Solar Sales Pty Ltd in Welshpool. As well as basing the chassis on the proven cyclone resistant Telstra framing system illustrated in Figure 23, for further security, all wiring was cabled through underground conduits were practical, and each set of panels allocated its own switch box for isolation and control purposes. All DC circuitry was connected by double-poled breakers to prevent arching when making and breaking circuits.

To provide a clear understanding of the principle of how solar panels are connected in series to increase voltage and then in parallel to increase current flow or amperage, a simplistic analogy of the manner in which an array of solar panels operates is given in the following analogy.

The reader is asked to consider the basic function of the "natural solar farmer"—the tree. A tree uses photosynthesis to convert simple minerals and elements into complex molecules. Using the analogy of a tree, consider the solar cells of each panel as leaves. The energy harvested by each cell, in the leaf, is directed into the leaf stem. The stems are connected to twigs and then larger and larger branches until they are all joined at the trunk.

The trunk is then connected to the root system, which is a structure in a similar fashion to the leaves and branches, which breaks up into smaller and smaller parts until ultimately the finest part of the root system is reached, the roots hairs—or in the case of solar-generated energy, the electrical appliance being powered by the energy collected by the solar/leaf cells, which are exposed to the sunlight.

The analogy is illustrated in Figure 29, which depicts a "natural" tree and the novelly of a constructed "solar tree." The "root system" of the "solar tree" is the electricity grid which is not depicted but rather left to the reader's imagination.

Figure 29:   A "natural" tree and its novelly constructed solar panel substitute. (Illustration: Moondust Design)

A pivotal consideration in configuring solar pv systems is that solar panels have a relatively low-voltage output yet the enabling equipment has a relatively high-voltage operating range. Solar panel output voltages range from around 12 V DC to around 65 V DC. Inverters, on the other hand, have minimum input voltages as low as 150 V DC ranging up to 600 V DC, with a preferred range to facilitate a "maximum power point tracking" (MPPT) function for optimum performance of between 230 and 500 V DC.[179]

It is, therefore, necessary to configure the solar panels into series and then into parallel strings to produce electricity at a voltage appropriate to the technical specifications of the inverter used to interface with the grid supply specifications. The "strings" must also be configured to operate within the ambient temperature specifications of the panels, while remaining within the amperage (current flow) constraints of the inverter.

The selected solar panels for the initial stage of the Solex project were General Electric GEPV-110 110 W photovoltaic modules, which were the most powerful available in Western Australia at that time. They were connected to three single-phase 5,000 W Fronius IG 60 high-voltage grid-tie inverters. Each 5000 W set of panels was connected to an inverter such that the entire configuration operated as a 15-kW three-phase unit connected to the Carnarvon Town Distribution grid operating at 240 V, 50 Hz, AC.

---

[179]   Fronius International GmbH, Fronius Australia, *Grid-connect Inverters* (2015) <http://www.fron ius.com/cps/rde/xchg/SID-3FCAF372-00042740/fronius_australia/hs.xsl/25_4961.htm#.VhCH9Yahfcc> at October 4, 2015.

The electrical circuit configuration is illustrated in Figure 30 below. The following section details the design parameters and estimate system performance considerations to produce an estimate of annual energy harvest. Those estimates will be compared to the actual energy production from the solar farm in Chapter 4.

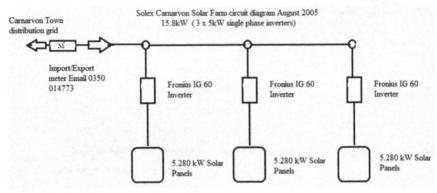

Figure 30:   Circuit diagram of the Solex project August 2005.

Design Parameters and System Performance Considerations

General Electric GEPV-110 Nominal Operating Cell Temperature (NOCT)

Variations in ambient air temperature are a critical factor in the operation of solar pv cells.

> As temperature increases, the open-circuit voltage decreases rapidly, while the short-circuit current increases slowly. Power output is voltage multiplied by current and so will decrease as well. As hot temperatures adversely affect power output, output from a pv array has to be calculated taking the temperature effects into consideration, i.e. derating an array's output based on the operating temperature conditions. Likewise, as cold temperatures can increase the power output due to the voltage increase, the maximum voltage threshold cannot exceed the inverter's ratings.[180]

Twidell and Weir suggest that "a module operating at 65°C (quite possible in a sunny desert environment) loses about 16 [per cent] of its nominal power."[181]

In order to assess the impact of variation of ambient air temperature, and therefore the variation of cell operating temperature and resultant power output performance, the impact of the variation of temperature to power output must first be estimated.

To establish a basis for the variation of cell operating temperature, a standard has been set as a nominal point of reference. That reference is referred to as the nominal

---

[180]   Geoff Stapleton and Susan Neill, *Grid-connected Solar Electric Systems: The Earthscan Expert Handbook for Planning Design and Installation* (2012) 54.
[181]   Twidell and Weir, above n 13, 223.

operating temperature (NOCT). Stapleton and Neill define the NOCT as being "[t]he temperature at which the cells in a pv module will operate under the following conditions: 800 W/m² irradiance, 20°C ambient temperature and a wind speed of 1 m/s."[182]

The NOCT is not to be confused with the concept of "Standard Test Conditions" (STC), which is defined as "[a] set of standard conditions under which pv cells are tested so that they can be compared. These conditions are 1000W/m² irradiance, air mass of 1.5 and a cell temperature of 25°C."[183] The STC provide a basis for the comparison of various makes and models of solar pv panels.

The following formula is used in this book to estimate the impact of temperature variation on the solar panels used in the construction of the Solex project.

$$T_m = (T_{NOCT} - 20)\ \frac{G_i}{800} + T_{amb}$$

where T_m is the cell temperature, G_i the irradiance, T_{amb} the ambient air temperature, and T_{NOCT} the nominal operating cell temperature.

Stapleton and Neill do not use a formula as above but, rather, simply add 25°C to the ambient air temperature to establish the maximum operating cell temperature and use the minimum air temperature as the minimum operating cell temperature.[184]

At the time of designing the electrical circuitry of the Solex project in 2004, the highest daily maximum temperature was 43.7°C.[185] It was, therefore, considered that 45°C to be the highest ambient air temperature that the pv cells would have to work in. Unfortunately, that maximum daily temperature was surpassed at 47.8°C on March 6, 2007, and again at 47.8°C on January 20, 2015.

In order for the solar array to operate within the ambient air temperature range in Carnarvon to ensure DC voltage output remained within the solar grid-tie inverter design constraints, the following calculations were made:

$$T_{max} = (T_{NOCT} - 20)\ \frac{G_i}{800} + T_{amb}$$

The data sheet for the GEPV-110 110 Watt photovoltaic module shows T_{NOCT} to be 45°C.[186]

---

[182] Stapleton and Neill, above n 185, 227.
[183] Ibid 229.
[184] Ibid 129.
[185] January 16, 1975, Australian Government Bureau of Meteorology, *Climate Data Online* (2015) <http://www.bom.gov.au/climate/data/> at February 9, 2015.
[186] Data specification solar pv module GEPV-110. GE Energy, *GEPV-110 110 Watt Photovoltaic Module* (2004).

Therefore, the maximum operating temperature that the cells would ever be expected to operate in was calculated as:

$$T_{max} = (45 - 20)\frac{1000}{800} + 45$$
$$= 76.25 \text{ °C}$$

By comparison, Stapleton and Neill would consider the maximum operating temperature to be

$$T_{max} = 45 + 25$$
$$= 70°C[187]$$

That comparison will be considered in the following calculations to ascertain if there is a discernible difference between the two methods and if there is a critical impact on safety margins or estimated performance.

The minimum operating temperature is the ambient air temperature at sunrise when the panels begin operation. The recorded lowest daily air temperature at Carnarvon was 2.4°C on July 25, 1979. While the daily minima are most likely to occur before dawn and not at sunrise, the differences are not sufficiently significant to affect the minimum cell temperature and, therefore, the maximum voltage for operational purposes.

Maximum String Voltage

The operating voltage parameters of the Fronius high voltage inverters model chosen, the Fronius IG 60 HV, for the maximum power point tracking range are 150 V DC to 400 V DC with a maximum input voltage of 530 V DC.[188] The open circuit voltage of the GEPV-110 solar panels is shown as 21.2 V DC in the manufacturer's specifications.[189]

Voltage Temperature Coefficient

The voltage temperature coefficient describes a module's change in the open circuit voltage (V_{oc}) per degree variation in temperature. The coefficient is described as a negative to indicate that voltage drops as cell operating temperature increases above the STCs under which the panel performance is characterized. The symbol β is used to signify the voltage temperature coefficient (mV/°C).

The following formula is used to calculate the effect of temperature on panel voltage to ascertain the maximum string voltage where V_{oc} refers to the module open

---

[187]   Stapleton and Neill, above n 185, 129.
[188]   Data specifications inverter model IG 60 HV, Fronius Australia, above n 179.
[189]   GE Energy, above n 191.

circuit voltage and V_{mp} refers to the maximum power point voltage of the panel. Both values are prescribed on the manufacturer's specifications of performance characteristics under STCs.

$$V_{max} = V_{oc} + \beta(T_{min} - 25°C)$$

For the GEPV-110 photovoltaic modules used on Stage 1, the maximum voltage of the solar panels was calculated as

$$V_{max} = 21.2 + (-0.08 \text{ V}(2.4 - 25)) = 23 \text{ V}$$

Therefore, the maximum number of panels in a series could not exceed 400/23 = 17.39 or 17 to the nearest lowest whole number for safety reasons.

Stapleton and Neill use a similar formula:

If the temperature is lower than 25°C:

$Voltage_{at\,x°c}$	$= Voltage_{at\,STC} + [\gamma_V \times (T_{x°c} - T_{stc})]$
where:	
$Voltage_{at\,x°c}$	= voltage at the specified temperature in volts;
$Voltage_{at\,STC}$	= voltage at STC, i.e., the rated voltage in volts;
γ_V	= voltage temperature coefficient in V/°C;
$T_{x°c}$	= cell temperature in °C;
T_{stc}	= temperature at STC in °C (i.e., 25°C).[190]

By substituting the values for the GEPV-110 110 W pv module as provided by the manufacturer's data sheet[191] and the forecast minimum ambient air temperatures for Carnarvon, the following calculation is made:

Voltage 2.4°C = 21.2 + [−0.08 × (2.4 − 25)]
= 23V

That is the identical result to the preceding formula, which gives a maximum number of 17 panels per string. The comparison indicates the use of varying symbols by researchers and designers to produce identical design parameters.

Minimum String Voltage

Once the upper parameter of panel numbers in series is established, the minimum number must be established to ensure the system continues to function at the highest expected ambient air, and therefore cell operating, temperature. The minimum voltage of the series of panels occurs when panels are at their hottest. Consideration must be given to the lowest operational voltage to ensure it does not fall below the operating range of the inverter—in the case of the Fronius IG60 HV, 150 V DC.

---

[190] Stapleton and Neill, above n 185, 130.
[191] GE Energy, above n 191.

To calculate the minimum number of panels to be used in series, the following formula is used:

$$V_{min} = V_{mp} + \beta(T_{max} - 25°)$$

For the GEPV-110 photovoltaic modules used on Stage 1, the minimum voltage of the solar panels was calculated as

$$V_{min} = 16.7 + (-0.08V(76.25 - 25)) = 12.6 \text{ V}$$

Therefore, the minimum number of panels in a series could not be less than 150/12.6 = 11.90 or 12 to the nearest highest whole number for safety reasons.

Once again, Stapleton and Neill use a similar formula:

If the temperature is higher than 25°C:
$$\text{Voltage}_{at\,x°C} = \text{Voltage}_{at\,STC} - [\gamma_V \times (T_{x°C} - T_{stc})]^{192}$$

By substituting the values for the GEPV-110 110 Watt photovoltaic module as provided by the manufacturer's data sheet[193] and the forecast maximum ambient air temperatures for Carnarvon, the following calculation is made.

$$\text{Voltage } 70°C = 16.7 + [-0.005 \times (70 - 25)] = 16.47 \text{ V}$$

Therefore, the minimum number of panels in a series could not be less than 150/16.47 = 9.10 or 10 to the nearest highest whole number for safety reasons.[194]

Stapleton and Neill's calculation produces a lower figure as their estimation of the maximum cell operating temperature is slightly lower as it ignores the variance between regional irradiance and the STC irradiance value of 800 W/m^2 and simply adds 25°C to the ambient air temperature. The first formula considers the cell temperature will rise a further 6.25°C, due to the higher irradiance level.

The discrepancy highlights the significance of impact of temperature on solar pv cell performance. Above 70°C, the impact in this case is equivalent to two panels or 220 W.

The chosen chassis structure as illustrated in Figures 23 and 24 created the requirement of even numbers for uniformity of construction, an upper and a lower string on each frame. Therefore, an even number of panels was required. Rows of six to provide a string of 12 as a minimum proved too small for practical purposes.

As the building material is stocked in 6 m lengths and the panels were 661 mm plus the width of a 20 mm clamp on either side, the nearest practical row number was eight to form series of 16 panels. The configuration produced a series minimum voltage of 201.6 V DC, or 263 V DC according to Stapleton and Neill's estimation, and a

---

[192]   Stapleton and Neill, above n 185, 129.
[193]   GE Energy, above n 191.
[194]   Stapleton and Neill, above n 185, 132.

series maximum voltage of 368 V DC and was well within the voltage operating parameters of the Fronius IG 60 inverters.

In addition to the effect of the variation of temperature from the standard testing condition of 20°C, losses caused by dirt and soiling, panel variations in manufacturing, shading, orientation, and tilt, cable losses, and inverter efficiencies reduce the anticipated system yield. Stapleton and Neill calculate the system yield by multiplying all of the derating factors together.

The energy output of the Solex project is metered. Therefore a detailed examination of anticipated system performance for "deeming" purposes is not essential as the concept of "deeming" RECS is outside the scope of this book.[195] However, to provide a basis for expected performance, the examination of the additional influencing factors affecting anticipated energy harvest is conducted here. The examination is pursuant to that detailed by Stapleton and Neill and accepted by the Australian Clean Energy Regulator for STC deeming purposes.[196]

The formula used in this estimation is:

$$(f_{tot}) = (f_{temp}) \times (f_{dirt}) \times (f_{mtol}) \times (f_{shad}) \times (f_{orien}) \times (f_{\Delta V}) \times (f_{inveff})$$

where:

(f_{tot}) = Total derating of all factors;
(f_{temp}) = temperature derating factor;
(f_{dirt}) = provision for dirt/soiling;
(f_{mtol}) = manufacturers tolerance;
(f_{shad}) = provision for shading;
(f_{orien}) = variation for orientation and tilt;
$(f_{\Delta V})$ = cable losses (volt drop);
(f_{inveff}) = inverter efficiency.[197]

The **temperature** derating factor for the GEPV-110 modules installed in Carnarvon is calculated as follows:

Cell temperature at the average ambient temperature of 27.27°C[198]

---

[195] "Deeming system performance" is the practice of setting an estimation of system performance for the creation of RECs, in particular STCs, which cannot be monitored for administrative and/or practical purposes. The Australian Clean Energy Regulator sets out precise guidelines by which STCs are "deemed" to be allocated according to anticipated energy harvest over the expected operating life of the system. Deeming of STCs was mentioned in Chapter 2 but is otherwise outside the scope of this book.
[196] Stapleton and Neill, above n 185, 142–143.
[197] Stapleton and Neill, above n 185, 142.
[198] Australian Government Bureau of Meteorology, above n 92.

$$T_{ave\,op} = \frac{(45-20)\underline{1000}}{800} + 27.27$$
$$= 58.52°C$$

Difference between the cell temperature and the STCs[199]:

$$T_{ave} = 58.52°C - 25°C$$
$$= 33.52°C$$

Temperature coefficient multiplied by the difference between cell temperature and STC

$$(f_{temp}) = 1 - (0.005^{200} \times 33.52)$$
$$= 1 - 0.1676$$
$$= 0.8324$$

The **dirt and soiling** derating factor for the GEPV-110 modules installed in Carnarvon is accepted in accordance with Stapleton and Neill's recommendation for low rainfall sites of:

$$(f_{dirt}) = 0.90^{201}$$

The **manufacturer's** tolerance derating factor for the GEPV-110 modules installed in Carnarvon given from the manufacturer's data sheet as:

$$(f_{mtol}) = 0.95^{202}$$

The **shading** derating factor for the GEPV-110 modules installed in Carnarvon is accepted in accordance with Stapleton and Neill's recommendation for north-facing arrays in full sun as being unaffected by shading; therefore:

$$(f_{shad}) = 1$$

The **variation for orientation and tilt** for the GEPV-110 modules installed in Carnarvon is accepted in accordance with Stapleton and Neill's suggestion that to use the system design guidelines for Clean Energy Council accredited designers. Discussions with architects revealed that an orientation of 10° east of north was favored by "eco-architects."

Architects considering reductions in household energy consumption through building design suggested that, by biasing the north/south axis of buildings slightly, east of north took advantage of morning temperatures being slightly lower than the ambient air temperature in the afternoon. This is caused by the duration of sunlight exposure on the atmosphere being the influencing factor on ambient air temperature. There

---

[199] Stapleton and Neill, above n 185, 138.
[200] GE Energy, above n 191.
[201] Stapleton and Neill, above n 185, 139.
[202] GE Energy, above n 191.

are of course other factors such as wind patterns and diurnal temperature variations of the land and sea. However, the architects considered that orientating the buildings as described had the effect of mitigating rising internal temperatures from exposure to the sun, particularly in summer.

As the objective of solar farms is to maximize solar energy harvest, it was decided to orientate the solar array in the opposite direction—that is, 10° to the west or an azimuth of 350°. Data provided by the Australian Clean Energy Council system design guidelines tables support that hypothesis.[203]

Data extracted from the guideline tables for annual daily irradiation on an inclined plane at an azimuth of 350° and an inclination of 20° for Alice Springs at latitude 23.7° south shows an average annual derating factor of 112.67 per cent. As Carnarvon's latitude is 24.9° south, the data for Alice Springs are considered acceptable; therefore, the derating factor for orientation and tilt for the Solex project is considered to be:

$$(f_{orien}) = 1.1267$$

The Australian standard for **cable losses** permits a maximum voltage drop of 5 per cent[204]; therefore:

$$(f_{\Delta V}) = 0.95$$

The efficiency for the Fronius IG 60 HV inverter installed in Carnarvon given from the manufacturer's data sheet is shown as 94.3 per cent[205]; therefore:

$$(f_{inveff}) = 0.943$$

"The total de-rating factor due to system losses is calculated by multiplying all the de-rating factors together."[206] By applying the total derating formula and substituting the abovementioned calculations for each derating factor, the following total expected system yield is calculated as follows:

$$
\begin{aligned}
(f_{tot}) \quad &= (f_{temp}) \times (f_{dirt}) \times (f_{mtol}) \times (f_{shad}) \times (f_{orien}) \times (f_{\Delta V}) \times (f_{inveff}) \\
&= 0.8324 \times 0.90 \times 0.95 \times 1 \times 1.1267 \times 0.95 \times 0.943 \\
&= 0.71835 \text{ or } 71.8 \text{ per cent}
\end{aligned}
$$

Therefore, the total system loss is estimated to be 28.2 per cent.

---

[203] Clean Energy Council, *Accredited Installer: Accreditation Guidelines* (2014) <http://www.solaraccreditation.com.au/installers/compliance-and-standards/accreditation-guidelines.html> at October 27, 2015.
[204] Stapleton and Neill, above n 185, 140.
[205] Data specifications inverter model IG 60 HV, Fronius Australia, above n 182.
[206] Stapleton and Neill, above n 185, 142.

According to Australian Bureau of Meteorology data, Carnarvon receives an average of 6.125 kWh/m²/day[207]; therefore, the annual irradiation level is determined as 2,235.6 kWh/m²/year.

Using this method, the total anticipated system yield for the 15.8-kW array is calculated as:

2,235.6 kWh/m²/year × 15.8 kW × 0.718 = 25,362 kWh/year.

This compares favorably with the estimated yield of 25,628 kWh prescribed by the multiplying factor of 1.622 MWh/kW installed in the *Renewable Energy (Electricity) Regulations 2001*.[208]

Chapter 4 reveals that the actual harvest achieved was considerably higher than these estimates. The Solex harvest is metered using a calibrated Email 0350 three-phase meter with an accuracy of ±2 per cent.[209] The meter is placed at the point of transfer to the Carnarvon Town site distribution network; therefore, the exported energy is obtained after all derating factors and losses considered earlier.

Panel layout of the frames was a matter of trial and error. The frames were manufactured and assembled on site; however, the actual layout of the frames was initially incorrectly designed; array was dismantled and reorganized into three uniform rows of three frames east to west.

While a "best estimate" was calculated to prevent shading between the rows at the summer and winter solstices, it is as much matter of luck as of good mathematics that barely 10 mm of shading between panel rows occurs at the height of the winter solstice. On June 21, the shadows cast by the arrays just strike the bottom of the panels arrayed to the rear. The shading in this case is immaterial to the performance of the rear arrays, but there is no "good clearance."

The calculation was simple geometry; however, as the panel angle, and, therefore, the overall height and the length of the resulting shadow vary according to the panel as described in Figure 24, provision had to be made for "the worst case" that is at the winter solstice, an panel angle of 45° from the horizontal.

Steel was purchased subject to a scope of works to complete nine frames as illustrated in Figures 23 and 24. It was fabricated on site by Fullarton and his son Andrew, with the assistance of Mr. Rick Freeman.

---

[207] Australian Bureau of Meteorolgy <http://www.bom.gov.au/jsp/ncc/cdio/weatherData/av?p_nccObsC ode=193&p_display_type=dailyDataFile&p_startYear=2003&p_c=-7398636&p_stn_num=006011>.

[208] *Renewable Energy (Electricity) Regulations 2001*, above n 142.

[209] Email from Craig Deetlefs Horizon Power Corporation to Alexander Robert Fullarton, October 29, 2015 (held by author); see also Australian Government: National Measurement Institute, *NMI M 6-1 Electricity Meters Part 1: Metrological and Technical Requirements* (2000) 4.

A sea-container was purchased to freight the solar panels and enabling equipment to Carnarvon. All materials were on site on February 3, 2005, and work commenced.

Anecdotally, Henry Ford required his automobile parts to be delivered to his factories in custom-designed crates. The crates were disassembled at the factory and had been designed to become parts for the automobile bodies. Similarly the sea-container, once stripped of its freight, became the inverter/transformer room. The concept has become standard practice in rural and remote Western Australia for other installations.

The work was hot, long, and hard. An additional privation was that, just as Carnarvon experiences general flooding roughly every 10 years at the beginning of each decade, so too does it generally experience a "wet" year (rainfall above average) every intervening 5th year; 2005 was no exception. The rainfall for May and June of that year experienced around three and a half times the annual average. In fact, at a recorded rainfall of 276 mm for those two months, the combined rainfall exceeded the annual mean average of 226 mm.

The impact was bogged vehicles in the Pindan clay that has virtually no structural capacity when soaked and the ever present determined and pestilent midge. To combat the annoying and irritating insects, fires were set up wind. It was preferable to work in the smoke than to put up with the biting insects. Extracting heavy machinery such as concrete trucks and heavy cranes became a practiced art. That phase was complete as the rain stopped, the insects vanished, and the track dried out.

The famous Australian poet "Banjo" Patterson once wrote "the Bush has friends to meet him and their kindly voices greet him." [210] The bard's romantic vision of "the Bush" may have been a little different had he ventured in the North West of Western Australia prior to compiling that prose.

Fabrication of the chassis commenced on February 3, 2015, and took Fullarton and his son Andrew 216 hours to complete. However, on April 7, it was realized that the site layout was wrong, and the chassis were removed and realigned into trenched footings by April 28. Panels were affixed to the chassis in strings of 16 and connected in groups of three to bring each AC phase to 5 kW connected to a Fronius IG60 HV inverter. The three phases were then connected to the grid as one single three-phase generation unit. Ultimately, the final solar pv capacity was a little over 15 kW at 15.8 kW. The installation was tested and commissioned by Western Power line staff on July 28, 2015.

---

[210] Andrew Barton (Banjo) Paterson, *Clancy of the Overflow* (1889).

The Official Opening

The name gives the impression that Solex is associated with solar or the sun. In truth, it was derived from a satirical label affixed to the meter cabinet by the electrical contract Mr Mick Millson. Mr Millson is of an Irish background and has a devilish sense of humor that the Irish are famed for. The label reads "SOLEX FULLOFIT DEVELOPMENTS." The implication is not complimentary but rather refers to the Australian ethos of telling tall stories or "yarn spinning" as previously described in the preface. A business name was needed nonetheless, and the label was the inspiration for the name "Solex."

The Solex Carnarvon Solar Farm was officially opened by the Hon Alan Carpenter, the Western Australian Minister for Energy, on October 6, 2005. The minister was kind enough to fit the opening of Western Australia's first privately owned solar farm at what could be considered very short notice in the Parliamentary world and the busy schedule of a State Government senior minister.

Opening functions are reasonably common and boring events wherein everyone says nice things about the project and declares them open. The Solex project had to be slightly different and the source of another interesting anecdote.

Fullarton had been a real estate agent for many years and, as part of advertizing and selling campaigns, is to provide the guest speaker with a gift to mark the occasion. As the project was a solar farm in the inhospitable weather conditions of Carnarvon in October, it was considered that a genuine "Bush" hat would be an appropriate gift. The archetypical Australian Akubra Cattleman style was chosen.

In the outback town of Carnarvon, the purchase of a Bush-hat is a fairly mundane and common commodity to purchase. However, as with all things of Solex, nothing was ever as easy as it first appears. Minister Carpenter's office was contacted to ascertain his head size.

The request was met with incredulity. It was simply not possible to get Minister Carpenter a hat. No one knew his head size and in any event such a hat did not exist. His wife was contacted, and she gave the same reply. An old friend who ran a "King-size Menswear" chain of shops was contacted. He had a good idea of what Minister Carpenter's head size would be. Alan Carpenter was well known in Western Australia as he had been a television journalist for many years prior to entering parliament and the old friend "had an eye" for menswear especially larger sizes. "Why he is a 64" was the reply "but I have no idea where you would get one."

The "hunt for the hat" began. Eventually, the manufacturer Akubra was sought and contacted.

The consultant was equally as confident of the head size as she had viewed Alan Carpenter on television many times. There were, as a matter of coincidence, two such hats in existence. They had been manufactured, for a customer from northern Queensland—a gray one and a brown one.

Akubra had intended to use the one the customer did not chose as a display model as it was expected that there would be no other demand for it in Australia. The hat arrived days before the ceremony, and everyone was sworn to secrecy as to the present.

The Minister was met at the airport and was full of apologies that he could not find a hat and would just have to risk sun burn. He opened the door of the vehicle and saw the hat on the seat. He pounced on it as a cat pounces on a mouse. Well, that present was certainly appreciated.

There was more to the tale yet to be told. Part way through the opening ceremonial speech, the Honourable Minister interrupted his speech with a tale of his own.

He came from rural stock in Albany in the south of Western Australia. From the age of around 12, all of his family and friends had good bush hats but Alan never got one. He would wait each birthday and Christmas for a hat but it never came. He later got married and thought perhaps his wife might get him such a present—but it never came.

His children grew to adulthood and he still held out hope for a hat for Christmas, birthday, or whatever. Years went passed, and he eventually gave up all hope of ever being the proud owner of an iconic Akubra. It then turned out that Lex Fullarton, given about a week's notice, had procured and produced such a present—the iconic Akubra Cattleman Bush hat.

He then mumbled "Maybe I should have him running my departments." The story of the hat became legendary.

Figure 31:   The Honourable Minister for Energy Mr Alan Carpenter and "The Hat." (Source: Karl Monaghan APP AAIPP [Associate Australian Institute of Professional Photography])

Figure 32:   Minister Alan Carpenter and Lex Fullarton, October 6, 2005. (Source: Karl Monaghan APP AAIPP [Associate Australian Institute of Professional Photography])

When the system was switched on, nothing perceptible happened. There was no noise or any other indication that the plant was operational except a couple of indicator lights and a glowing led screen display. A person was heard to say "I knew this wouldn't work." To which Mr Carpenter replied "it sure is, I can see the meter were it is feeding in." He then asked very quietly "how long have you had this running?" "Oh about a couple of months" was the reply. "You didn't think I would have got you up here if I didn't know it was going to work did you?"—Environmentally friendly energy generation had arrived in Carnarvon.

Stage 2—30.2-kW Solar Array Extension

The success of the 15.8 kW encouraged an extension of the solar pv capacity based on the original construction with variation to the solar panels and enabling equipment used. Sunpower SPR-210 210 W panels and a Fronius IG 400 30 kW central inverter were chosen as the solar panels had developed in the intervening two years and Fronius had released the central three-phase inverter into Australia.

Mr Tamun Davidson of Solar Sales designed the electrical layout in accordance with the data constraints shown below.

Design Parameters and System Performance Considerations

This section will apply the same design factor considerations as calculated for the 15.8 kW array; however, the calculations will avoid duplication of common factors, such as shading, cable losses and orientation, and tilt, and calculate only those variables specific to the Sunpower pv solar panels and the Fronius inverter model IG 400.

Sunpower SPR-210 Normal Operating Cell Temperature (NOCT)

$$T_m = (T_{NOCT} - 20) \frac{G_i}{800} + T_{amb}$$

The data sheet for the Sunpower SPR-210 210 Watt pv module shows T_{NOCT} to be 46°C.[211]

Therefore, the maximum operating temperature that the cells would ever be expected to operate in was calculated as:

$$T_m = (46 - 20) \frac{1000}{800} + 45$$
$$= 77.5°C$$

---

[211] Data specification solar pv module SPR-210-BLK-U. Sunpower Corporation, *210 Solar Panel 210 Watt Photovoltaic Module* (2009).

Maximum String Voltage

The operating voltage parameters of the Fronius high-voltage inverters model chosen, and the Fronius IG 400 CE HV for the maximum power point tracking range are 210 V DC to 420 V DC with a maximum input voltage of 530 V DC.[212] The open-circuit voltage of the Sunpower SPR-210 solar panels is shown as 47.7 V DC in the manufacturer's specifications.

Voltage Temperature Coefficient

For the Sunpower SPR-210 pv modules, the maximum voltage of the solar panels was calculated as

$$V_{max} = 47.7 + (-0.1368 \text{ V}(2.4 - 25)) = 50.79 \text{ V}$$

Therefore, the maximum number of panels in a series could not exceed 420/50.79 = 8.26 or 8 to the nearest lowest whole number for safety reasons.

Minimum String Voltage

For the Sunpower SPR-210 pv modules used on Stage 2, the minimum voltage of the solar panels was calculated as

$$V_{min} = 40.0 + (-0.1368 \text{ V}(65 - 25)) = 34.53 \text{ V}$$

Therefore, the minimum number of panels in a series could not be less than 210/34.53 = 6.08 or 7 to the nearest highest whole number for safety reasons.

As with Stage 1, the chosen chassis structure as illustrated in Figures 23 and 24 created the requirement of even numbers for uniformity of construction, an upper and a lower string on each frame. However, in this case, as the chassis were capable of carrying an entire string on one row, an even number of panels was not essentially required. Rows of seven were possible but proved too small for practical purposes as more chassis would be required. Therefore, the same configuration of eight panels to the row as used in Stage 1 was selected.

Other Design Parameters and Derating Factors

The formula used for calculating total system yield is as for Stage 1:

$$(f_{tot}) = (f_{temp}) \times (f_{dirt}) \times (f_{mtol}) \times (f_{shad}) \times (f_{orien}) \times (f_{\Delta V}) \times (f_{inveff})$$

The **temperature** derating factor for the Sunpower SPR-210 210 Watt pv module installed in Carnarvon is calculated as follows:

---

[212] Data specifications central inverter model IG 400, Fronius Australia, above n 184.

Cell temperature at the average ambient temperature of 27.27°C[213]

$$T_{ave\ op} = (46 - 20)\frac{1000}{800} + 27.27$$
$$= 59.77°C$$

Difference between the cell temperature and the STCs:[214]

$$T_{ave} = 59.77°C - 25°C$$
$$= 34.77°C$$

Temperature coefficient multiplied by the difference between cell temperature and STC:

$$(f_{temp}) = 1 - (0.0038^{215} \times 34.77)$$
$$= 1 - 0.1321$$
$$= 0.8679$$

The **dirt and soiling** is as for the GEPV-110 module array detailed earlier:

$$(f_{dirt}) = 0.90^{216}$$

The **manufacturer's** tolerance derating factor for the Sunpower SPR-210 modules installed in Carnarvon given from the manufacturer's data sheet as:

$$(f_{mtol}) = 0.95^{217}$$

The **shading** is as for the GEPV-110 module array detailed earlier:

$$(f_{shad}) = 1$$

The **variation for orientation and tilt** is as for the GEPV-110 module array detailed earlier:

$$(f_{orien}) = 1.1267$$

The Australian standard for **cable losses** is as for the GEPV-110 module array detailed earlier:

$$(f_{\Delta V}) = 0.95$$

The efficiency for the Fronius IG 60 HV inverter installed in Carnarvon given from the manufacturer's data sheet is shown as 94.3 per cent[218;] therefore:

$$(f_{inveff}) = 0.943$$

---

[213]   Australian Government Bureau of Meteorology, above n 92.
[214]   Stapleton and Neill, above n 185, 138.
[215]   Sunpower , above n 216.
[216]   Stapleton and Neill, above n 185, 139.
[217]   Sunpower, above n 216.
[218]   Data specifications inverter model IG 60 HV, Fronius Australia, above n 184.

"The total de-rating factor due to system losses is calculated by multiplying all the de-rating factors together."[219] By applying the total derating formula and substituting the abovementioned calculations for each derating factor, the total expected system yield is calculated as follows:

$$(f_{tot}) = (f_{temp}) \times (f_{dirt}) \times (f_{mtol}) \times (f_{shad}) \times (f_{orien}) \times (f_{\Delta V}) \times (f_{inveff})$$
$$= 0.8679 \times 0.90 \times 0.95 \times 1 \times 1.1267 \times 0.95 \times 0.943$$
$$= 0.749 \text{ or } 74.9 \text{ per cent}$$

Therefore, the total system loss is estimated to be 25.1 per cent.

The annual irradiation level is that calculated earlier as 2,235.6 kWh/m²/year. Using this method, the total anticipated system yield for the 30.2 kW array is calculated as:

2,235.6 kWh/m²/year × 30.2 kW × 0.749 = 50,569 kWh/year.

The extension was expected to increase the solar farm annual harvest to 75,931 kWh, or nearly 76 MW per year. Actual output is given in Chapter 4.

The benefit of the experience of constructing Stage 1 resulted in Stage 2 being completed by Fullarton and his son-in-law Mr Logan Trigg in just 82 days, compared to the 175 days that Stage 1 had taken though Stage 2 is roughly the same physical dimension.

It is also noted that technology of connecting the solar panels had improved. The requirement to individually connect each junction box was substituted by a multicontact electrical "plug" system, a MC3 fitting, constructed in manufacture, which removed the laborious task of "hard wire" cable connections.

The Sunpower 210 solar panel was not only more powerful and easier to install than the General Electric GEPV-110 but also introduced a different approach to cell connectivity. In an attempt to improve efficiency, Sunpower introduced the concept of connecting the cells from the rear. The small wires on the face of "traditional" solar panels are not visible in the Sunpower series.

That technological advancement also had an unintended consequence. It was discovered that a charge of static electricity was built up on the face of the panel over a period of time. Eventually, the electron flow from the panel became "neutralized." Performance fell to the point where the panels ceased to function entirely. That problem was overcome by "earthing" the positive connection at the inverter. The earth functions as a "drain" and permits the panel to function in a normal fashion. The "positive earth" is shown in the circuitry diagram in Figure 33.

---

[219] Stapleton and Neill, above n 185, 142.

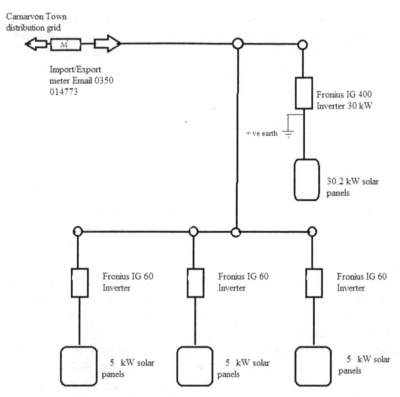

Carnarvon Town
distribution grid

Import/Export
meter Email 0350
014773

+ ve earth

Fronius IG 400
Inverter 30 kW

30.2 kW solar
panels

Fronius IG 60
Inverter

Fronius IG 60
Inverter

Fronius IG 60
Inverter

5 kW solar
panels

5 kW solar
panels

5 kW solar
panels

Figure 33:   Circuit diagram of the Solex project: August 2007.

Stage 3—The Interogation of a 5-kW Westwind Wind Generator

As indicated earlier, the Solex project is a "family business" and Fullarton's second son, Simon, was involved in this stage. He was also a tradesperson employed in the later construction of the ice-works—Stage 4.

It is trite to observe that a solar farm is only operational during daylight hours, and therefore, from a commercial perspective, the enabling equipment has a 50 per cent redundancy. It only operates for half of the 24-hour day. Mr Laughton-Smith, of Horizon Power, highlighted that when he noted the total duration of energy production for Stage 1, shortly after the commissioning of Stage 2.

He noted that total operation for Stage 1, at that point, was around 8,000 hours. That appeared reasonable as there are 8,760 hours in a year. However, the farm had been in operation for two years, not one. That observation highlighted the fact that electricity production ceased during the hours of darkness—that is half of the day.

117

It was suggested that the solar pv enabling equipment may operate regardless of the source of DC energy supply and that "Even a man on a bicycle might be able to generate energy acceptable to operate the system." That was a reference to the "pedal" radio sets once used in Australia for Outback communications.

Those statements instituted investigation into integrating a source of wind energy with the solar enabling equipment during the hours of darkness. Investigations were conducted to establish the integration of a 5-kW wind turbine with a 5-kW Fronius IG 60. Staff at Fronius suggested that, provided the DC energy supply was maintained within the input voltage parameters of the inverters, then the inverter operation should be possible.

Westwind engineers considered it was possible to manufacture a wind turbine engineered to enable wind energy to be fed into the grid, by way of the Fronius solar enabling equipment. An adaption was made to a standard 120-V DC Westwind turbine, to produce 240-V DC generation, which interfaced with a Fronius IG 60 inverter.

The innovation provided an alternative energy source to the solar panels, which provided for electricity production to be continued beyond the hours of daylight. It was considered that Carnarvon's solar/wind profile particularly suited this arrangement as the region's significant wind strengths are experienced before sunrise and continue well after sunset.

An additional innovation was the fabrication of a "folding gin pole" to the standard Westwind designed mast. Owing to the perennial threat of tropical cyclones, described earlier in this chapter, the arrangement of the standard design of an 18-m Westwind stayed mast left the integral gin pole, essential for raising and lowering the mast, in an upright position. It was realized that a 6-m mast would be threatened by cyclonic winds.

A solution was to redesign the mast such that not only did the mast lower to the ground for safety but so too was the gin pole able to be further lowered to lie on the upper side of the mast. A far smaller 2-m removable gin pole was designed to fit into the main gin pole in order to raise and lower that section.

Figure 34 shows wind turbine number 3 being lowered prior to tropical cyclone Olwyn in March 2015. The gin pole to give leverage for raising and lowering the mast is clearly visible at the right hand side of the photograph. The two masts behind are already lowered showing their gin poles lying on the mast section to reduce potential damage caused by the impact of cyclonic winds.

Figure 34:   Lowering turbine 3. (Source: Photograph taken by Alexander R Fullarton, 2015)

The worker standing to one side gives a perspective of the dimensions of these "small generation systems." The 15 kW of wind turbines occupy approximately twice the area of the 53 kW of solar installations behind them.

Figure 35 shows the first 5-kW Westwind turbine in position in the background of part of the array of Stage 1 of the solar panels it integrated with.

Figure 35:   Westwind 5-kW turbine and solar panels. (Source: Photograph taken by Alexander R Fullarton, 2008)

The circuit diagram shown in Figure 36 indicates the switching system that permitted DC energy input from either wind or solar. The innovation did not increase the generational capacity of the solar farm but rather permitted a choice of energy sources, particularly beyond daylight hours.

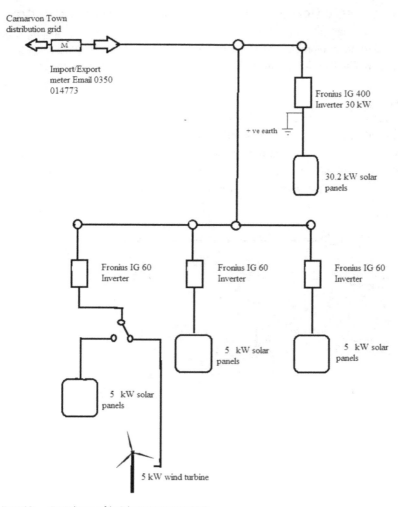

Figure 36: Circuit diagram of the Solex project: August 2008.

It was anticipated that a 5-kW wind turbine located in Carnarvon would generate 9.9 MWh per annum. The actual results are given in Chapter 4. The innovation also permitted data to be collected to ascertain the comparative effectiveness of wind and solar generation systems. From that data, the economic comparatives could be established. That analysis will also be provided in Chapter 4.

In September 2009, an opportunity arose to purchase an additional Fronius central inverter IG 400. That additional inverter capacity permitted the three single-phase sections of the GEPV-110 operating as three single-phase systems into the one three-phase central inverter.

This additional equipment was not considered a development stage as it did not, directly, increase the operational capacity of the solar arrays. However, it did provide for an increase in generational capacity by replacement of more powerful solar panels later, in Stage 7.

That additional equipment permitted the wind turbine to operate in concert with the solar arrays and effectively increased the overall generation capacity by 5 kW, as well as permitting direct comparison of the two energy sources. Figure 37 illustrates the full integration of the wind turbine and development of the Solex project into a solar/wind farm. Chapter 4 reveals the results of that comparison.

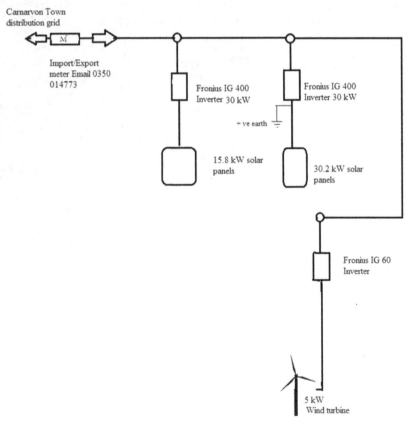

Figure 37:   Circuit diagram of the Solex project: September 2008.

Stage 4—The Construction and Integration of an ice-Works, 2009

The concept of "value adding" to the solar farm's renewable energy production was conceived after a visit to Tasmania. It was observed that much of the state's considerable hydroelectric resource was consumed in the refining of ore. The importation of raw materials and processing by way of sophisticated infrastructure was beyond the resources of the Solex project but not purchasing potable scheme water and ice-making machines.

An ice-works, consisting of a factory building, ice-manufacturing machines, freezer, and ancillary plant, was constructed to produce around 160 tonne of cubed ice per annum for the commercial market. The plant was to demonstrate and alternative use for alternative energy and intended to compete directly with ice manufacturers using fossil fuel to manufacture store and transport their product.

The plant was design without the advantage of any experience or knowledge of the design, layout, or operation of a commercial ice-works. Therefore, the Solex ice-works is a little unique in its manufacture, processing, and storage of product.

In order to ensure the ice-works "load profile" of energy use coincides with daylight hours and the harvest of solar energy daylight switches have been installed to control the ice machines. The daylight switches are connected to the "bin full" sensors on the machines and turn the machines off during the hours of darkness.

Another adaption to the manufacturing system is the use of customized fibreglass bins placed under the two Manitowoc 1800 ice-manufacturing machines. As illustrated in the Figure 38, the ice machines have been placed on commercial pallet racking, rather than atop a stainless steel ice bin provided by the manufacturer.

The "standard" storage system requires the ice to be manually shovelled out through a top access to the bin below. That packaging system was considered laborious and not really suited to an industrial process. Therefore, in place of the manufacturers' fixed bins, the customized bins were designed and manufactured locally.

The key feature of the customized bins is that they are mobile and are removed for packaging purposes. They are then jacked into position and the contents raked out from an access chute in the bottom of the bins.

The ice bins are an example of the composite use of technology and the variety of the origin of manufacture of the plant used in the Solex project. Smaller articles, such as pallet jacks, are manufactured in China, the ice machines are from the USA, the ice bins were fabricated locally in Carnarvon, and so on.

Pride of place is a Matthiesen Volumetric Bagger, which was manufactured to Australian electrical standards in San Antonio, Texas. Despite being able to operate from Australian 240-V 50-Hz AC electricity, it has its USA compliance designation 120-V 60-

Hz AC labeling affixed to its frame. The construction of the machine to Australian standards was a matter of negotiation with the manufacturer, as demand for such a product in Australia was limited.

Figure 38:   The Solex ice manufacturing system. (Source: Photograph taken by Alexander R Fullarton, 2013)

The solar ice-works consumes around one-third of the solar/wind farm's annual production, thereby continuing to offset other energy stake holders and unavoidable fossil fuel energy costs of the Water Corporation's town reticulation system.

The ability to manufacture ice locally provides an independent and reliable source of an essential product in times of emergency—floods, fires, and cyclones. Local manufacture of ice also displaces the costs of road transportation of the product, some 500 km from the nearest alternative point of manufacture, as well as the fossil-fueled energy used to manufacture and store the product.

It is acknowledged that the production of Solex ice production is not entirely fossil fuel free as the supply of water is carried out by external entities are powered by fossil-fueled energy sources. In addition, some energy is drawn from the fossil fuel-based utility grid from time to time for overnight storage and ice manufacture. However, the use of solar energy in ice production greatly reduces the impact of fossil fuel combustion, as well as demonstrating that the two energy sources; dispersed renewable energy and centralized fossil-fueled power stations, can function co-operatively rather than competitively.

The ice-works also incorporates a 4.6-kW roof-mounted solar pv array, and the circuit diagram depicting the addition of that installation on the ice-works building is shown in Figure 39.

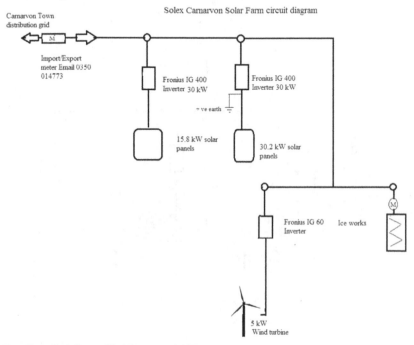

Figure 39:   Circuit diagram of the Solex project: July 2009.

The economic and environmental outcomes of the solar ice production will be examined in Chapter 4.

Stage 5—The Integration of a further two 5-kW Wind Generators, 2010

The production level of the wind turbine was far less than anticipated from previous investigations of wind energy resource for Carnarvon, and that estimated by the Clean Energy Regulator for the purposes of deeming carbon credits (LGCs). However, despite limited success of the initial turbine it was decided to "complete the set" by installing two further Westwind 5-kW wind turbines into the solar farm enabling equipment. They were connected to the two remaining AC phases to produce a three-phase system. Figure 40 shows the generation circuitry as at January 2010.

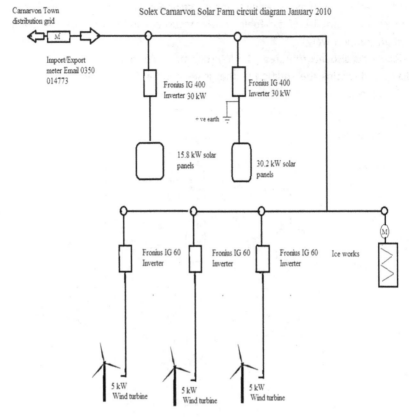

Figure 40: Circuit diagram of the Solex project: January 2010.

In 2012, a 4.6-kW roof-mounted solar pv system was installed on the ice-works. That system contributes around one-third of the plant's energy requirements. Figure 41 illustrates the growing complexity of the Solex project from a simple energy harvest and generation of electricity to the Carnarvon town distribution system redirecting some of that energy to provide energy to "add value to the raw material." Note the dual system of meters on the ice-works to account for energy into the ice-works from the solar farm as well as its own supply.

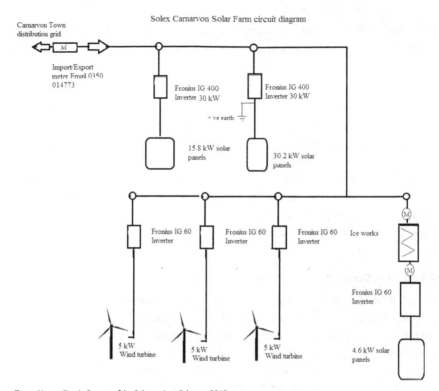

Figure 41:   Circuit diagram of the Solex project: February 2012.

In the event excess energy is produced, the meter monitoring the 4.6-kW solar pv system runs forward to account for the energy produced, the meter from the solar farm into the ice-works runs backward as the energy flows out of the building, and the main meter monitors the energy flow to the distribution grid. In that way, only energy consumed by the ice-works is accounted for and any surplus is monitored as it leaves the Solex project for external sale.

Using the method suggested by Stapleton and Neill examined previously and substituting the manufacturers' tolerances and other variables calculated for the solar farm Stages 1 and 2, the total anticipated system yield for the 4.6-kW array is calculated as:

$$(f_{tot}) \quad = (f_{temp}) \times (f_{dirt}) \times (f_{mtol}) \times (f_{shad}) \times (f_{orien}) \times (f_{AV}) \times (f_{inveff})$$
$$= 0.8545^{220} \times 0.90 \times 0.979^{221} \times 1 \times 1.1267 \times 0.95 \times 0.943$$

---

[220]   Chint Solar (Zhejiang) Co., Ltd, *Datasheet Crystalline PV Modules CHSM 6610P series Astronergy 235wp* (2012).

[221]   Ibid.

= 0.75994 or 76.0 per cent

Given the annual irradiation level previously calculated above as 2,235.6 kWh/m²/year, annual energy harvest is expected to be:

2,235.6 kWh/m²/year × 4.6 kW × 0.76 = 7,816 kWh/year.

For comparative analysis, the actual energy harvest data will be presented in Chapter 4.

In addition to the portion of "self-generation" provided by the solar pv installation on the ice-works roof, daylight switches have been installed to switch the ice machines off during hours of darkness to ensure they only operated during daylight hours when solar energy is being harvested.

That feature enforces a shift in the load profile (energy consumption pattern) to match the period when solar energy is being produced. If necessary, they can be overridden to continue ice production beyond daylight hours. Unfortunately, that encompasses using a fossil-fueled–sourced energy, to ensure production matches demand from time to time.

Stage 6—The Construction of a Farmhouse, 2013

In 2013, a farmhouse was constructed on the land in order to reduce costs in time and fuel traveling to the farm from a residence in the town of Carnarvon. The home was constructed in Perth, some 900 km away and transported to a prepared section of the land.

Later, in January 2015, the farmhouse also had a 5.4-kW roof-mounted solar pv system installed to further increase the Solex project's energy generation capacity. The circuit diagram depicting the addition of the farmhouse and its roof-mounted solar pv system is shown in Figure 42.

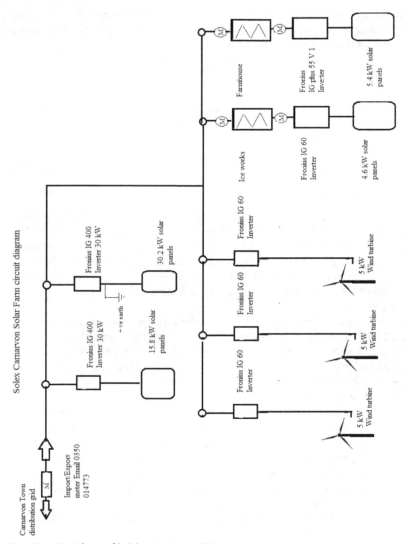

Figure 42:   Circuit diagram of the Solex project: January 2015

The metering arrangement is as for the ice-works with a meter at the supply point to the building for energy from the solar farm and a meter incorporated into the Fronius IG Plus V1 5.5-kW capacity inverter. Excess energy from the building reverses the supply meter in the same way to prevent double counting.

By substituting the manufacturers' tolerances for the Astronergy 270W modules, and other variables calculated for the solar farm Stages 1 and 2, the total anticipated system yield for the 5.4-kW array is calculated as:

$$(f_{tot}) = (f_{temp}) \times (f_{dirt}) \times (f_{mtol}) \times (f_{shad}) \times (f_{orien}) \times (f_{\Delta V}) \times (f_{inveff})$$
$$= 0.8369^{222} \times 0.90 \times 0.97^{223} \times 1 \times 1.1267 \times 0.95 \times 0.957$$
$$= 0.7484 \text{ or } 74.8 \text{ per cent}$$

Given the annual irradiation level previously calculated above as 2,235.6 $kWh/m^2/year$, annual energy harvest is expected to be:

2,235.6 $kWh/m^2/year \times 5.4$ kW $\times 0.748 = 9,030$ kWh/year.

For comparative analysis, the actual energy harvest data will be presented in Chapter 4.

Stage 7—The Replacement of GE 110 W Solar Panels with ASTRONERGY 270W

It was decided to replace the GEPV-110 110W panels with the more powerful Astronergy 270W panels to increase output and to utilize the redundant capacity of the number two central inverter, the 30-kW Fronius IG 400. Eight of the nine chassis containing Stage 1 were striped of 16 GEPV-110 modules each (14.08 kW) and had nine Astronergy 270W modules replaced on them (19.44 kW). That increase capacity by 5.36 kW from 15.8 kW to 21.16 kW without requiring further enabling equipment of infrastructure.

One of the chassis, which is on the roof of the inverter room, was left with the original 16 GEPV-110W modules to investigate the concept of mismatching solar panels by type. It was found that the total voltage (V_{oc}) of the 16-panel string of GEPV-110W modules was within 2.35 per cent of the voltage of the nine panel string of Astronergy 270W modules. That is well within the manufacturers' tolerances of both manufacturers of ±5 per cent.

The inverter has not revealed any faults or disturbances to output from expected performances. It is, therefore, suggested that, provided total voltages and power flows (amperages) are matched within the power tolerances of manufacturers, the use of differing makes and models of solar panels is acceptable. However, it is recommended that professional engineering advice be sought before designing such a combination of modules as this trial cannot be considered a reliable, or compliant, basis generally. It merely demonstrates that it can be done.

The anticipated performance and annual energy harvest based on previous calculations is:

2,235.6 $kWh/m^2/year \times 19.44$ kW $\times 0.748 = 32,508$ kWh/year
Plus 2,235.6 $kWh/m^2/year \times 1.76$ kW $\times 0.718 = \underline{2,825}$ kWh/year
Total anticipated annual harvest for the upgrade = 35,333 kWh/year

---

[222] Chint Solar (Zhejiang) Co., Ltd, *Datasheet Crystalline PV Modules CHSM 6612P series Astronergy 270wp* (2010).
[223] Ibid.

It is planned to install further   modules to this array to fully exploit the capacity of
the Fronius IG 400, which is 30 kW. The circuit diagram of the Solex project including
the upgrade of Stage 1 is shown in Figure 43. Figure 43 includes the intended replacement of 5-kW solar panel arrays in substitution of the wind turbines decommissioned
in March 2015.

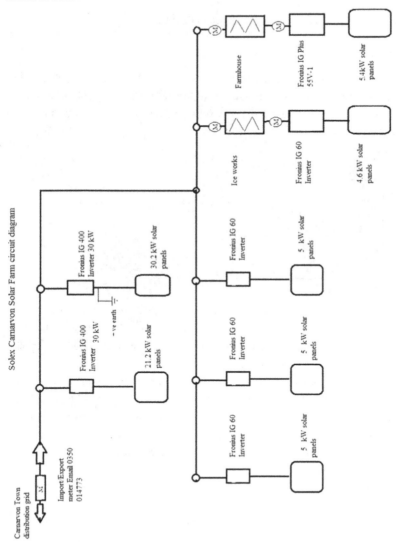

Figure 43:   Circuit diagram of the Solex project: March 2015

Stage 8—The Removal of the Wind Generators, 2015

The wind turbines were lowered in March 2015 in preparation for tropical cyclone Olwyn. Data were collated over five years and due to the less than expected harvest, particularly in relation to the comparative harvest of an equal capacity solar pv system, they will be replaced by 15 kW of solar arrays. For comparative analysis, the actual energy harvest data will be presented in Chapter 4.

3.6   SUMMARY

Ultimately, a total generational capacity of 61.4 kW of solar pv was installed on the Solex project. The performance calculations carried out in this chapter indicate a total expected energy harvest of around 132 MWh/a, being the aggregate of all energy systems as shown in Table 5.

Stage	Annual energy harvest (kWh)
1	25,628
2	50,569
3	9,369
4	7,816
5	19,800
6	9,030
7[224]	9,705
Total	**131,944**

Table 5:   Projected annual energy harvest.

Included in the is 15 kW of wind generation capacity, which was trialled, however, as discussed in Chapter 4, the results were disappointing and that will be replaced by solar pv renewable energy harvesting technology. That replacement will be considered as Stage 8, which is yet to be constructed.

While it is acknowledged that the energy harvest from the wind source was considerably lower than forecast, an important factor must be considered—the climatic factors peculiar to the region in which the comparison is made. Carnarvon has a reasonably predictable wind energy resource; however, in the dry, desert-like conditions of clear skies, strong sunlight, and resulting high-insolation levels of the near 25° south latitude, it is hardly surprising that solar energy is a far more efficient source of renewable energy.

This chapter has covered the inception of the Solex project and tracked its progress of staged development, from the acquisition of desert land, to the construction of a

---

[224]   Increase in capacity to stage 1.

small demonstration solar energy trial plant, and finally to the development of a commercially productive venture. The Solex project introduced the concept of "alternative use for alternative energy" by directing some of its energy harvest into on-farm manufacturing.

The intent of the project was to demonstrate economic enterprises can be conducted in an environmental and socially sustainable manner, even on land once considered to be "waste land." Fullarton saw this land not as "waste country" but rather "country wasted" and set out to prove it.

Chapter 4 will detail actual performances of the Solex project components and compare and contrast that data with the production estimates calculated in this chapter. It also looks at other aspects of the triple bottom line in the savings of fossil fuel consumption and avoided pollution of greenhouse gas emissions. It shows the savings of those emissions by way of LGC/RECs created and traded over the 10 years of operation of the project from 2005 to 2015.

CHAPTER 4—ECONOMIC AND
ENVIRONMENTAL OUTCOMES

"That's one small step for man, one giant leap for Mankind."

Neil Alden Armstrong, NASA Astronaut (1930–2012).
After having stepped on the Moon July 21, 1969

4.1   INTRODUCTION

This chapter presents the findings of the trials and the outcomes of the stages of development detailed in the previous chapter. It compares the estimated costs and expected production and fiscal returns estimated in Chapter 3, with those actually achieved over the 10-year period: 2005–2015.

It also attempts to provide explanations for the discrepancies between the scientifically based estimates and the actual energy harvests. However, this book is a case study of one solar pv installation in one location and not a valid basis to establish reliable scientific conclusions.

Data sourced from the Solex project may be insufficient to provide reliable scientific evidence to establish firm base lines for future applications. However, it will provide a point of reference for further study into solar energy harvest in the northwest of Western Australia and Carnarvon in particular. It may also assist in further research for solar pv energy harvest in the tropical regions of the planet—latitudes below 25°.

The environmental impacts of the Solex project to improve the natural vegetation and encourage endemic fauna to reestablish communities on the land, and its environs will be examined as additional methods of reducing greenhouse gas emissions from the atmosphere.

For convenient comparative analysis this chapter is structured in the same sequence of development stages used in Chapter 3.

4.2   ECONOMIC CONSIDERATIONS

Stage 1—15.8-kW Solar Array

As detailed in Chapter 3, the budget for the construction of Stage 1—the 15-kW solar installation, was $174,500. Though the final constructed cost increased by 14.2 per cent to $199,339, the final capacity also increased by 5.6 per cent to 15.84 kW. On a unit cost basis, the final constructed unit cost rose slightly from the estimated $11.63/W to $12.58/W.

During the preparations for the establishment of the first stage, it was found that authorities often considered scientific estimates and projections as irrefutably accurate and based their evaluations accordingly. In the 10 years of operation, Solex data clearly demonstrate those presumptions to be inaccurate.

An increase in energy harvest of over 25 per cent on "formal" projections was attained. The actual generation for the first year was 32,059 kWh, or an average of nearly 88 kWh per day. That compared to the 65-kWh per day output anticipated by SEDO and the 70-kWh per day as calculated in Chapter 3.

That result is attributed, in part, from higher than standard insolation levels of 1,000 Wm^2 being experienced in the Carnarvon region, which is up to 1,200 Wm^2. The additional energy harvest is also due to the advantage of seasonal solar tracking capabilities, which allow the solar pv panels to face south for the period of the summer solstice (in the southern hemisphere). The ability to track the sun on a seasonal basis is primarily enabled by the chassis design, which permits that maneuver.

This outcome demonstrates the necessity to support or refute scientific modeling, with data collated from performance of trials held in the field. However, without reliable field data to refute the scientifically based estimates and projections used by relevant authorities to determine grants, rebates and other subsidies the "standard" economic performance analysis had to be considered as valid.

That approach to economic analysis was very frustrating and persisted for many years until actual performance data were collated. The attitudes of officers within State authorities toward administrative procedures and statutory compliance often led to considerable time delays in administrative processes and funding.

Initial frustrations have led to amusement in the longer term. Often administrative concerns, raised by individuals within government organizations, were addressed repetitively due to turn over of staff. Philosophical perspectives are considered to be social factors, rather than economic or environmental considerations, and are examined further in Chapter 5, which looks at social influencing factors.

Stage 2—30.2-kW Solar PV Array Extension, 2007

The success of Stage 1 provided reliable data as to the impact of possible disturbances to the electricity grid caused by the solar pv installation. The two-year trial suggested that the solar installation it provided an element of stabilization to the distribution system rather than negatively impacting on it.

The higher than expected energy harvest also indicated a better economic forecast than that considered by Western Power in 2004. In addition, there had also been an increase in electricity prices. Until then, electricity tariffs had been artificially suppressed by the Western Australian government for many years.

The positive economic result and technical data collated influenced the decision to expand the "trial" by adding a further 30 kW of generation capacity. Energy harvest increased accordingly with 84,963 kWh produced in 2007–2008 compared to the 75,931 kWh estimated in Chapter 3 for both stages.

Energy harvest for subsequent years for the two stages is:

2008–2009 82,488 kWh
2009–2010 93,321 kWh
2010–2011 84,425 kWh (loss of harvest due to flooding)
2011–2012 94,951 kWh
2012–2013 89,748 kWh
2013–2014 83,552 kWh (delay in adjusting panels, Stage 1 u/s for a month)
2014–2015 86,034 kWh (TC Olwyn)

As with any other form of farming, the vagrancies of the weather affect solar farming. It was found that, apart from tropical cyclones and floods, which cannot be avoided, seasonal adjustments to tilt are critical at the latitude of 25°. It is also acknowledged that additional capacity of 5.4 kW was added in September 2014. It will not be until 2016 that the impact of that improvement will be quantified.

The loss of harvest caused by the impact of Tropical Cyclone Olwyn will have been compensated for by the higher capacity installed to give the appearance of consistency in energy harvest.

The data also tend to indicate there is no apparent deterioration in system efficiency other than from enabling equipment (inverter) failures. Regular monitoring of the installation is essential but not onerous. It has also been found that washing panels is unnecessary provided tilt angle is always greater than 12.5°. Despite the low rainfall of apparently dry desert climate of Carnarvon, there is sufficient rain to prevent significant dust build ups.

Stages 3 and 5: 5-kW Westwind Wind Generators, 2008 and 2010

Energy harvest from the Westwind wind generators was extremely disappointing. Scientific projections based on wind data for Carnarvon indicated that around 9.4 MWh of energy would be harvested from a 5-kW wind generator. However, data collated by Solex over a five-year period reveals that an average annual harvest was less than half of that at 3.8 MWh.

The initial advantage of lower construction costs was eroded by the rapid decline in solar pv panel costs, and ultimately, the poorer than expected energy production resulted in a decision to remove the wind generators in favor of solar pv panel arrays.

A further consideration was the area of land required to stay and raise and lower the masts. The three generators, with a total generation capacity of 15 kW and an average annual energy harvest of 11.4 MWh, occupy an area that could reasonably house 100 kW of solar panel arrays. That solar pv capacity could have generated as much as 200 MWh per annum, using the Solex annual tracking chassis.

However, the trial was worth conducting to establish actual field data to support of refute the scientific estimates. The Solex data suggest that, for the location of Carnarvon, solar pv is an economically superior method of renewable energy harvest to wind. The ratios will vary from place to place depending on a number of climatic factors. This book suggests that it would be a mistake to apply the findings of the Solex data to all locations on Earth.

The trial also investigated integrating wind generation with the solar pv systems. That trial revealed that wind generators could be supported by the solar inverters and had an interesting outcome. Apart from the weakness of requiring a comparatively large area of land, the wind generators were able to supplement solar generation, particularly in the hours of darkness.

The following discussion provides an indication of how wind and solar can function in a collaborative manner. As demand for energy generally falls during the hours of darkness as load profiles tend to the pattern of human activity, the reduced energy harvest of Carnarvon's wind resource becomes less critical. Therefore, wind may be able to support overnight "base loads" if required.

The comparative analysis begins with the energy generation of solar pv arrays. It then looks at the pattern of energy harvest from a wind generator of the same size, and finally, it reveals how the combined system functions over a three-day period. It was conducted in roughly average winds for Carnarvon during the "trade wind" period of the month of February.

Figure 44 shows the reasonably even parabolic trajectory of the path of the sun in the clear skies of Carnarvon on a typical day. The slight dip in at around 13:15 followed by a relatively constant output is the effect of the sea breeze. The rapid fall and return to the "normal trajectory" of output is caused by shadowing of large trees to the west of the array.

Note also that zenith at Carnarvon is 12:30 as it is west of the central point of time zone "Hotel" (UTC + 8 hours). The slight orientation of 10° west, which distorts the parabola toward afternoon harvest is also evident.

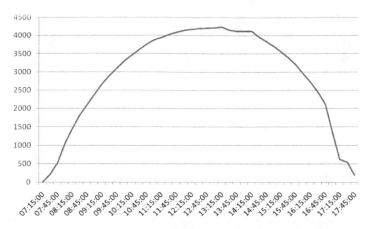

Figure 44: Typical daily solar energy production. (Source: Solex Fronius IG60 HV datalogger[225])

The following two figures indicate the effectiveness of the combined solar/wind system. Figure 45 shows the erratic nature of the wind over a three-day hour period. Wind gusts give rise to high variations. Figure 46 shows the combined use of solar and wind by switching from solar during daylight hours to wind overnight, over a similar three-day period.

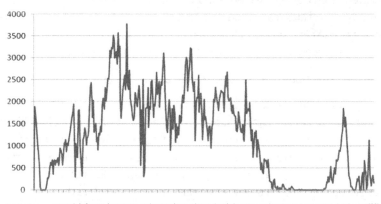

Figure 45: Typical daily wind energy production (over three days). (Source: Solex Fronius IG60 HV datalogger[226])

[225] Data provided by Solex Carnarvon Solar Farm IG 60 datalogger 5kWp solar plant July 12, 2005.
[226] Data provided by Solex Carnarvon Solar Farm IG 60 datalogger 5kW wind turbine February 10–12, 2009.

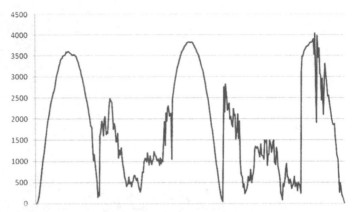

Figure 46:   Typical combination solar/wind energy production (over three days). (Source: Solex Fronius IG60 HV datalogger[227])

When Figure 44 is compared to Figure 45, the uniformity of solar pv harvest compared with the vagrancies of the wind generators is obvious. Energy production of wind is very intermittent and generally poor when the capacity of the wind installations is considered. On occasions during the trial period, August 2008 to February 2009, daily production from a 5-kW solar pv installation was up to seven times that of a 5-kW wind generator.

However, when run as a dual system, with wind sourced during the hours of darkness, the wind provides reasonable support to overnight loads. Given that demands for electricity also diminish at night, it appears solar/wind combinations could provide very convenient support to fossil-fueled energy installations generally.

Stage 4—The Construction and Integration of an ice-Works, 2009

This section examines the construction and integration of an ice-works in the Solex project to demonstrate a practical and economic example of using renewable energy in an industrial environment. As mentioned in Chapter 3, the ice-works has a roof-mounted solar pv array to supplement its energy consumption. Discussed further under the heading Stage 6, that solar array is an east/west "split" system.

The development of the ice-works was particularly challenging. The proponent had no knowledge of ice industry but only an elementary understanding of its manufacture, storage, and transport. There were no immediate ice-works within the region to emulate; therefore, design, construction, manufacturing, and marketing was largely a matter of trial and error.

---

227   Data provided by Solex Carnarvon Solar Farm IG 60 datalogger 5kW solar/wind turbine February 10–12, 2009.

Simple matters, such as a source of packaging materials, became logistical challenges as the only plastic bag manufacturer that could be found at the time was in Victoria. A method of bagging had to be implemented. Even sealing the bags to a commercial standard was challenging before suitable and effective sealing machines could be sourced.

The general public, and the sole distributer, Mr Dean Rowe of Caltex Starmart Roadhouse Carnarvon, were extremely patient with the new product as faults and weaknesses were progressively overcome. Without that support, Solex Solar Ice would never have been successfully promoted.

It was noted that general weaknesses in the production and marketing system were considered by most as part of the process of introducing an "environmentally preferred" product. It is believed that the general public tended to be more tolerant toward the development of an environmentally friendly product than they might otherwise have been.

Solex Solar Ice was marketed as being able to "cool down your drinks without warming the planet." It developed into a serious substitute product for the fossil-fueled alternative.

By the end of 2015, the Solex ice-works had produced just over 760 tonnes of ice with a market value of nearly $430,000. That may be a very humble economic achievement, compared to major industrialized manufacturers, but there are also considerable extrinsic economic savings.

Those economic savings are by way of fossil fuel displaced as well as environmental benefits of reduced greenhouse gas emissions generated by the alternative of manufacture and transportation of a substitute product over some 500 km. Figure 47 shows the pattern of ice production from the Solex ice-works and illustrates its move into the "mainstream" market.

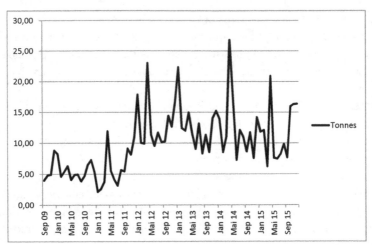

Figure 47:   Monthly solar ice production: 2009–2015.

One of the main intangible benefits is the volume of greenhouse gas emissions avoided by this innovation. Solex data collated over the seven-year period of operation indicate the savings in energy costs and greenhouse gas emissions to manufacture and transport the ice to Carnarvon. They reveal the avoided costs that would have been incurred had the product been manufactured and transported using fossil fuel.

It is estimated that approximately 95 tonnes of diesel and 205 tonnes of greenhouse gas emissions in manufacture and a further 9.5 tonnes of diesel and 20.9 tonnes of greenhouse gas emissions in transport were displaced.[228]

Solex data reveal around 316 MWh of energy has been consumed in the production and storage of roughly 760 tonnes of ice. Findings from that data indicate that, in the Solex ice-works, 1 kg of ice requires 415 Wh or 0.415 kWh of energy. By applying that finding to estimate avoided greenhouse gas emissions, using the ratios discussed in Chapter 2, the production of 1 kg of ice consumes 124.5 mL of diesel and produces 269.7 mg of greenhouse gases.

The following comparative matrix is suggested as an example of converting energy consumption to units of production experienced in the Solex project's ice production environment:

Units of ice production = units of energy = units of fuel consumed = units of greenhouse gas emissions

---

[228]   These figures are based on Solex data, which indicate 1 kWh of electricity is required to manufacture and store 2.41 kg of ice; the average road train carries 75,000 L of fuel and consumes 500 mL/km. Emissions are assumed to be 650 mg/kWh.

1 kg Solex solar ice = 414.97 Wh electricity = 124.45 mL diesel = 269.64 mg GHG emissions
1,000 Wh electricity = 2.41 kg Solex solar ice = 300 mL diesel = 650 mg GHG emissions
1,000 ml diesel = 8.03 kg Solex solar ice = 3,333.33 Wh electricity = 2,166 mg GHG emissions
1,000 mg GHG emissions = 3.71 kg Solex solar ice = 1,539.42 Wh electricity = 461.54 mL diesel

The overall renewable energy production, diesel fuel offset, and pollution avoided by the entire Solex project are examined later in this section.

An important environmental factor influencing the manufacture of Solex solar ice is the cyclical process of freezing and thawing water to ice to water, which ultimately has limited, or entirely no, detrimental impact on the natural water cycle and the broader environment.

Diagramatically, the environmental impact of Solex ice production is produced in Figure 48.

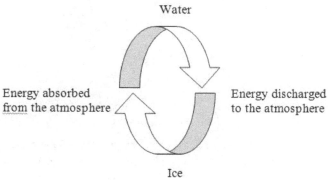

Figure 48:   The water–ice energy cycle.

Heat energy is drawn from the water, changing the water to its solid state—ice; the energy is discharged to the atmosphere as heat. In the reverse process, heat energy is absorbed from the atmosphere, by way of cooling the substances required to be cooled—foodstuffs, beverages, and the like; and the ice returns to its liquid state. The water then continues its passage through the natural water cycle.

The process does not reduce solar energy absorbed by the planet but it does displace the combustion of fossil fuel and associated atmospheric pollution.

From a broader perspective, the Solex ice production water cycle is illustrated in Figure 49:

- Water is pumped from the Gascoyne River through the Water Corporation's town water reticulation system;
- Potable water is frozen using standard ice manufacturing machines and freezers powered from the solar/wind farm;
- The ice is distributed to consumers, melts, and returns to the natural environment.

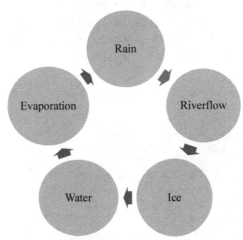

Figure 49:   Solar water–ice water cycle.

The ice-works consumes just a little short of 50 per cent of the Solex project's renewable energy generation but contributes 54 per cent of its revenue. At the same time, despite a relatively consistent supply of electricity to the Carnarvon Town electricity distribution system, falling purchase prices paid for renewable energy in Carnarvon, have seen the value of the "raw" electricity fall from 12.3 per cent of total revenue in 2010, to just 5.7 per cent in 2015.

The construction of the ice-works has, therefore, become essential to the economic existence of the Solex project. As with many "primary products," the revenue generated from the supply of raw materials is a small portion to the value of the "finished product." The balance of the Solex project's revenue is generated from renewable energy credit trading and consultations to the solar pv industry generally.

Stage 6—The Construction of a Farmhouse, 2013

As outlined in Chapter 3, a farmhouse was constructed on the Solex project to reduce fossil fuel consumption by way of travel. It incorporated a 5.4-kW roof-mounted solar pv array. Estimated annual energy harvest calculated in Chapter 3 indicates that some 9 MWh could be attained. The actual harvest of 8,450 kWh for 346 days of operation gives an indication of the accuracy of that calculation, being a mere 1.28 per cent variation.

The extremely small variation appears to give strong support for the accuracy of the predicted outcome, based on scientific theory. However, as illustrated in Figure 50, the array is not a single-planed, north-facing structure. As with the solar pv installation on the ice-works, the farmhouse array is a "split" system with 2.7 kW on each side of the roof ridge line facing east-west at a differential angle of 28°, being 14° pitch (tilt) on each side.

Initially, this method of solar panel array structures was resisted by designers. They feared departure from the "normal" single plane would have a detrimental effect on energy supplies and inverter performance. Subsequently, investigations by manufacturers have proven those fears to be unfounded, and the enabling equipment is able to cope with the unequal energy flows from the "split" array.

The outcome is that, depending on the size of the differential of planar angles, a solar panel array can be supported by enabling equipment of down to two thirds of the power of the total output capacity of the array as at no time does the array output its full capacity as a monoplane array structure. That is, a solar panel array of 6 kW can be supported by an inverter of only 4.5 kW. This structure has resulted in great cost savings and a reduction in constructed cost, which in turn permits higher rates of revenue return.

Figure 50:   Solex farmhouse 5.4-kW solar pv array. (Source: Photograph taken by Alexander R Fullarton, 2015)

The trial supports the design principle that, at latitude 25°, the east-west configuration not only provides an energy harvest within the predicted tolerances of a north-facing "standard" single-plane array but also incorporates a tolerance for the variation in azimuth of sunrise and sunset discussed in Chapter 3. It is further noted that, despite a total solar panel capacity of the array being 5.4 kW, the peak output recorded is only 5.215 kW, some 96.57 per cent of the inverter tolerance and well within manufacturer's specifications.

The configuration, which is particularly suited to buildings with north/south roof alignments, is favored as it provides higher energy harvests in summer when ice/energy demands are at their highest.

While it is suggested that, at latitudes of less than 25°, designers may be confident of using this configuration for roof-mounted solar arrays, the configuration is only suitable for inverters with multiple power point tracking (MPPT) capabilities.

Explanation of the function of MPPT inverter capabilities operate is beyond the scope of this book. Further information on the subject can be found at Fronius International GmbH Internet website.[229]

In addition to energy harvest from the solar pv installation, rainwater is also harvested from the farmhouse roof. The 345 m^2 roof area can collect nearly 80 kL of rainwater in an average year. At the tariff for water supplied by the water utility to the Solex project in 2015, that is a value of nearly $480 per annum. That is roughly equivalent value of a tonne of ice manufactured by the ice-works. Rainwater tanks installed at the farmhouse have storage capacity for 18.5 tonnes of rainwater, nearly a quarter of the average annual rainfall.

Stage 7—The Replacement of GE 110W Solar Panels with ASTRONERGY 270W Panels

The 15.8-kW array was increased to 21.2 kW in September 2014 by exchanging 128 GE 110W solar panels of the original array with 72 Astronergy 270W at a cost of $13,860 ($0.71/W). There was no recovery value from the technologically obsolete GE 110W panels. An alternative use for the panels such as charging 12-V systems for recreational use is being investigated.

The cost of the upgrade was relatively inexpensive; however, the increase in actual energy harvested was ultimately only around half of that initially estimated. No satisfactory explanation has been established for that. Further investigations will be made in 2016 to suggest why that has occurred.

---

[229] Fronius international GmbH, Fronius Australia, inverter with multiple mpp trackers: requirements and state of the art solutions (2013) < http://www.fronius.com/cps/rde/xbcr/sid-1148ef51-e86faada/froni us_international/se_ta_inverter_with_multiple_mpp_trackers_en_320367_snapshot.pdf> at December 13, 2015.

Stage 8—The Replacement of the Wind Generators, 2015

As at the end of 2015, this stage of the project was yet to be carried out. The turbines were lowered and will be dismantled and replaced by the equivalent generation capacity of solar pv panels. Each set of 5-kW panels will be integrated into the existing 5-kW Fronius IG 60 inverters that serviced the 5-kW wind turbines. It is expected that development will increase energy harvest for that stage by a factor of 5–6.

The following section looks at the impact of the Solex project in terms of fossil fuel displaced and the consequential reduction of associated greenhouse gas emissions that would otherwise have been discharged to the atmosphere.

4.3   ECONOMIC VIABILITY

Electricity Generation

The records of the Solex project reveal to the end of the financial year ended June 2015, it had generated electricity to the value of $173,354, of which $137,456 had been sold to the State Utility at ruling rates. The remainder was consumed by the ice-works. It was valued at the same tariff as the State Utility to express a true opportunity cost.

Most of the costs associated with the operation of the solar farm were minimal, other than the costs of insurance, finance, and capital depreciation. The costs of insurance required under the power purchase agreement with the State Utility are around $600 per calendar month or $7,200 per annum.

That is, for liability insurance to indemnify, the utility is case of injury to persons entering the solar farm. To date, an insurer has not been found to insure the solar panels against damage or destruction.

Finance costs of around 9 per cent on the constructed cost of the solar pv and wind installations after rebates is $31,000 per annum, and depreciation at a flat rate of 10 per cent over the 10-year period is $67,848 per annum.

Therefore, the resultant annual cash profit of $6,545 before interest and depreciation amounts to an accounting loss of $92,303 per annum. Mr Varvell's 2003 prediction that the project was economically unviable is supported by these financial results, though his financial assumptions may have been somewhat pessimistic.

Despite the comparatively minimal operational costs of solar pv energy generation the capital cost, which persists for many accounting periods, has made the Solex project fiscally unviable. Alternatively, fossil-fueled–powered energy generation systems have been far cheaper to install, even though they face the ongoing burdens of continual fuel costs and greenhouse gas emissions.

The abovementioned extract of accounts recognizes a primary weakness in the overall fiscal analysis in that the entire capital cost was depreciated over a 10-year term. Solex data reveal that there appears to be little perceptible decline in energy harvest from the solar pv panels as previously anticipated and that all equipment appears well serviced and operational.

Should operations continue for another 15 years, as anticipated by Mr Varvell, then the annual loss would be reduced by $67,848. It is also reasonable to assume that alternative energy costs will rise, thus increasing gross revenue and reducing the loss. In addition, declining capital costs, chiefly attributable to economies of scale in the production and installation of solar pv systems in the future, will greatly alter those fiscal dynamics.

Another influencing factor on the volume of gross revenue to the Solex project, from the sale of electricity, is the tariff for the sale of electricity to the State utility—Horizon Power. As discussed in Chapter 2, prevailing governmental policies and attitudes toward renewable energy in Australia influence the negotiated sale price of the Solex project's electricity.

This book suggests that, despite the appearance of being a profitable enterprise, Horizon Power is not profitable. The key point of discrepancy is focused on the accounting policies adopted by the corporation. The financial statements disclose an annual "Tariff Equalization Fund" payment made by the State to Horizon Power, which is recorded as "general revenue" in the financial statements.[230] In 2014, that subsidy was $209m and $136m in 2015.[231] This book argues that that payment is not "general revenue" but rather a taxpayer-funded subsidy to meet the annual operating losses incurred by the enterprise in its operations.

According to the financial statements, Horizon Power disclosed net profits as being $81.6m in 2014 and $38m in 2015. However, if the taxpayer-funded subsidy is removed from the accounts, they would actually disclose losses of $127.4m for 2014 and $98m for 2015.

In fairness, Note 31 to the financial report recognizes the importance of the subsidy to its financial position. Note 31 states:

> A significant portion of Horizon Power's revenue is derived from the Tariff Equalisation Fund (TEF). Western Power pays money into the Tariff Equalisation Fund in amounts determined by the Treasurer and the Minister for Energy. This money is released to Horizon Power as determined by the Treasurer. Horizon Power is dependent on the sufficient and timely flow of these

---

[230] Regional Power Corporation trading as Horizon Power, *Financial Statements for the year ended 30 June 2015*, (2015) 2.
[231] Ibid.

funds to remain solvent. Horizon Power began receiving revenue from the Tariff Equalisation Fund from October 2006.[232]

The impact of how Horizon Power's financial performance affects Solex's tariff rate is reflected by that economic focus. In 2015, Solex's contract tariff rate with the State utility was below that first proposed in 2005. It is noted that, until mid-2015, the tariff rate was on parity with the State utility's electricity tariff for the sale of its electricity. However, detailed examination of the financial processes and accounting operations, of the Western Australian electrical Tariff Equalization Fund is outside of the scope of this book.

As discussed in Chapter 2, prevailing political philosophies dominate TBL influences. In 2005, the ALP was the governing political party. As its attitudes toward society and the environment dominated over economic expediency, Solex received a favorable status when negotiating electricity sale prices to the utility. In 2015, when contracts were renegotiated, economic interests dominated utility focus of purchasing tariff rates. That change in focus was in part attributable to the economically dominated Liberal philosophy.

Therefore, Solex not only had to compete with fossil-fueled generation costs but rather subsidized fossil-fueled generation costs. Horizon Power's focus had changed from giving support to dispersed embedded solar pv installations (under the ALP government) to defraying its economic losses (under the Liberal government). Further discussion of community responses to the Solex project is conducted in Chapter 5.

However, it is noted here that the change in government political philosophy was illustrated in 2010, when restrictions were placed on the growth of community-based embedded solar pv installations by Horizon Power.

It is also noted that political philosophies can be influenced by the community. In late 2015, the Western Australian Government committed to supporting a further roll out of solar pv installations. It announced that it would provide a 2-MW solar storage and management system to Horizon Power for the purpose of extending the integration of further community-based solar pv systems into the Carnarvon electrical distribution system.

The supply of electricity has been supported by taxpayer contributions throughout the industrial age despite the resource being so critical to society and industry. This book suggests that, if the true cost of electricity was imposed upon industry and society, that impost would be prohibitive. The provision of energy to the community illustrates the necessity for essential community services to remain under some form

---

[232]   Ibid, 62.

of taxpayer funding and be functions of government regardless of economic expedience.

Additional revenue for the Solex project was sourced from the sale of carbon credits (RECs/LGCs). Seven hundred and twenty-two LGCs were created over the 10-year period. An additional 146 LGCs were deemed for the purposes of allocating the energy source into a category for wind as distinct from photovoltaic or solar. The LGCs were sold for as high as $75 each, but the average sale price was much less at $45 for most of the period.

Therefore, income from the sale of carbon credits amounted to a further $40,000 or $4,000 per annum. From 2015 onward, it is reasonable to expect that, given the tax-effective value of LGCs is $92.86 as calculated in Chapter 2, and Australia is yet to construct sufficient renewable energy installations for the next decade, the floor price of LGCs is likely to remain above $70 for some time.

Another negative revenue impact on the creation of LGCs for the Solex project was the deeming of LGCs for the purposes of categorizing the energy source. Of the 146 LGCs deemed in 2008 and 2010 and allocated to wind generation, only 58 were actually generated. In order to rectify the overestimation and claim the correct number of LGCs, some 88 LGCs had to be compensated for from the solar pv harvest.

That shortfall in anticipated wind generation returns resulted in another $800 per annum revenue loss, which will not reoccur in the forthcoming years.

Therefore, the next 15 years should reduce the revenue loss to $16,205 per annum rather than the $88,000 per annum loss experienced to date. Therefore, it is reasonably certain that, while revenue losses will not reach those envisaged by Mr Varvell in 2003, the fiscal returns from generating electricity will never be a profitable operation for the Solex project.

Ice Manufacture

An unintended consequence of the decline in tariff for the sale of "raw" energy is that energy "opportunity" cost to the ice-works has fallen commensurately. Despite Solex producing ice with a market value of $430,000 as disclosed earlier, the inclusion of goods and service tax reduces the revenue to Solex to just over $380,000 for the six-and-a-half–year period to the end of 2015. Therefore, a net revenue return of around $18,000 per annum is expected, given operational expenses are around $40,000 per annum.

However, given the initial two years of sales were just $22,000 per annum, in the remaining years, sales have risen to around $80,000 per annum. Therefore, with modest expansion, the Solex project may be reasonably profitable.

Up until the end of 2015, the Solex project had survived two floods and a tropical cyclone. As with any farming venture, it is subject to the vagrancies of weather and climate. While the project itself suffered no damage from any of those events, the damage to distributers and the community generally had severe financial impacts on fiscal returns. As with many primary production enterprises in Australia, Solex is supported by income from other activities.

In common with the State utility, Solex as at the end of 2015, "is dependent on the sufficient and timely flow of these funds to remain solvent,"[233] particularly as it faces the additional burden of reduced electricity prices for the sale of its "raw" electricity.

It is also interesting to note that a fiscal beneficiary of the Solex project is not Solex itself but rather the savings in fossil fuel imposts. Should the value of diesel displaced by the renewable energy source be brought to account, at $1 L, it would conservatively have the value of around $360,000. The savings to reducing greenhouse gas emissions are un-calculable. The State and the environment benefit from those savings as they are on no economic benefit to Solex, other than through the creation and sale of LGCs.

4.4   ENVIRONMENTAL CONSIDERATIONS

Atmospheric CO_2 Reductions

As stated in Chapter 2, the Solex project's prime directive is to reduce greenhouse gas emissions by substituting non-polluting renewable energy generation to lower fossil fuel consumption, as illustrated in Figure 8. This section reveals the outcome of that aim.

The following analysis is based on the scientific premise discussed in Chapter 2, that around 300 mL of fuel oil or "diesel fuel equivalent" is required to produce 1 kW of electrical energy.[234] Furthermore, it also considers that the process of combusting that fossil fuel emits just over twice its uncombusted volume in pollutant gases to the atmosphere.[235]

Figure 51 suggests a simplistic but practical daily activity as an example of the ratio of fossil fuel combusted in electrical consumption and the resultant impact on greenhouse gas emissions generated in doing so.

The illustration considers the use of a 1,000-Watt vacuum cleaner being operated for an hour. That operation consumes 1 kWh of electricity. Ignoring losses incurred by inefficiencies in generation and transmission, that "unit" requires 300 mL, or

---

[233]   Ibid.
[234]   Arceivala and Asolekar, above n 43.
[235]   Ibid.

roughly a Coca-Cola can full, of diesel to generate the kilowatt-hour. In doing so, roughly, the same volume of air is consumed to produce around two Coca-Cola cans of emissions.

Used for 1 hour   Consumes one of these   Produces two of these

Figure 51:   Greenhouse gas emissions from using a vacuum cleaner. (Illustration: Moondust Design)

Table 6 quantifies the Solex project's displacement of fossil fuel and accompanying greenhouse gas emissions as its contribution to reducing atmospheric CO_2.

Stage	Annual energy harvest				
	Estimated (kWh)	Actual (average) (kWh)	Value[236] ($)	Diesel displaced (L)	Emissions avoided (kg)
1	25,628	31,170	5,629	9,351	20,260
2	50,569	57,146	10,320	17,144	37,145
3 wind	9,369	3,838	693	1,151	2,495
4	7,816	9,005	1,626	2,701	5,853
5 wind	19,800	7,677	1,386	2,303	4,990
6	9,030	8,914	1,610	2,674	5,794
7[237]	9,705	4,404[238]	795	1,321	2,863
Total	**131,944**	**122,154**	**22,059**	**36,645**	**79,400**

Table 6:   Solex annual energy harvest by stage.

Table 6 shows that the Solex Carnarvon Solar Farm directly displaces over 36 tonnes of fossil fuel per annum and nearly 80 tonnes of greenhouse emissions. Those figures do not include the additional fuel and emissions displaced, which would have also been required to transport and store the fuel that has been displaced.

---

[236]   The value of electricity is the contracted purchase rate plus the value of the REC at the time of writing. There have been considerable variations to both, over the period of operation.

[237]   Increase in capacity to stage 1.

[238]   This figure has been calculated by comparing average monthly production for the entire solar farm before the panel replacement to that after June 2013 and removing the production data of the farmhouse installation. Investigations will commence in 2016 to establish the accuracy of this figure.

To provide an overall understanding of energy harvest of the Solex project, Figure 52 shows the pattern of renewable energy generated, fossil fuel, and greenhouse gas emissions avoided by the Solex project since it began operations.

Over the 10-year period, it has displaced over 360 tonnes of fossil fuel[239] and around 790 tonnes of greenhouse gas emissions avoided. It is estimated that these figures will at least double over the life of the project.

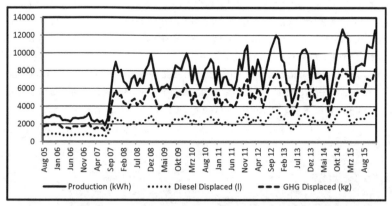

Figure 52:   The Solex project renewable energy production: 2005–2015.

This section has considered the impacts on the atmosphere; the following section looks as the terrestrial influences of the Solex project on the land it occupies.

Physical Environmental Impacts

As shown in Figure 8, in Chapter 2, planting carbon absorbing vegetation is one of the suggested methods of reducing existing atmospheric CO_2. Twidell and Weir suggest that

> [E]ach person obtains metabolic energy for 24 h from reaction with oxygen derived from about 15–30 m² of leaf area. Thus in temperate regions, one person's annual bodily oxygen intake is provided by approximately one large tree. In the tropics such a tree would provide metabolic energy for about three people. Industrial, transport and domestic fuel consumption require far more oxygen per person, e.g. about 100 trees per person in the USA, about 60 in Europe, about 20 in much of the developing world.[240]

A secondary objective to the Solex project was to revegetate the land to mitigate some of the physical environmental damage caused by the British Colonists in the early 20th century as discussed in Chapter 2.

---

[239]   For comparison, the average road train carries around 70 tonnes of fuel.
[240]   Twidel and Weir, above n 13, 329.

However, though land restoration was limited by the desert-like environment and climate, some rehabilitation steps have been taken. It is important to note that, while watering of trees took place for the first two to three years to establish them, revegetation has relied on natural rainfall and flood events. No general irrigation program has taken place.

By taking advantage of a series of better than average rainfall years, and in the belief that 2000 would be a "flood year" as outlined in Chapter 3, the land was ploughed in 1999 to break the hard clay surface in an effort to re-establish native vegetation. Grass clippings from local caravan parks and tree loppings were then scattered across the soil surface.

2000 was indeed a "flood year"; the nearby Gascoyne River over topped and flooded the entire area. Ten years later, in 2009, that action had resulted in a good covering of vegetation and grasses as illustrated in the photograph below in Figure 53.

Figure 53:   A view along Gascoyne Location 42 (now Lot 100): January 2009. (Source: Photograph taken by Alexander R Fullarton, 2009)

The same area of land is depicted in a photograph taken in 1994, in Figure 19, in Chapter 3, for comparison as evidence to the effectiveness of the limited land restoration work carried out.

The action to revegetate the land was most fortunate. Not only was 2010 a "flood year" as indicated by previous records discussed in Chapter 3, but it was the highest ever recorded in Carnarvon.

Solex project installations had been constructed at a level considered to be well above "flood reach," and coupled with the vegetation encouraged over the previous year, no flood damage occurred to the Solex project overall. The 2010 flood levels experienced on the Solex project provided an opportunity to evaluate anticipated future flood levels and provide guidance for future construction building levels.

In addition, a method was trialled to plant eucalypt trees into the hard clay soil with minimal irrigation. It was found that, by plant saplings around 1 m deep into an excavation of roughly the same diameter and backfilled with light sand, the trees became established trees within around 24 months of regular watering. Ten years later, the eucalyptus seedlings removed from the Gascoyne riverbed, and certain annihilation by the next river flow, are established trees some 10 m in height.

By the end of 2015, some 56 eucalyptus trees of a variety of species have been successfully established. It had been decided to trial a variety of Australia's great range of eucalypts with limited success. Tuarts, which are mighty trees in Western Australia's Great Southern region, flourished for a short time and grew quickly.

However, they did not withstand the ravages of damaging insects endemic to the Gascoyne Region. Likewise, swamp gums and ironbarks from New South Wales and Jarrah trees and others from far more clement clines did not survive. However, the local white or "ghost" gums, and some exotic red-flowering gums, have become established and attract a range of bird life and insects.

The following color plate exhibits examples of flowering eucalypts and birds and insects, which are now common-place on the land containing the Solex project.

The white-winged fairy wren is found in the drier parts of central Australia; from central Queensland and South Australia across to Western Australia. One or more males in a social group grow brightly coloured plumage during the breeding season. The female is smaller than the male and is sandy-brown with light-blue tail feathers. In spring and summer, groups of white-winged fairy wrens consist of a brightly colored older male accompanied by small, inconspicuous brown birds, many of which are also male.

The Red-capped Flowering Gum Eucalyptus erythrycorys or "Illyarrie" gets its common name from the interesting ribbed red caps which drop off to reveal an unusual four-sided flower with clusters of golden stamens. The trunk is white and leaves are scythe shaped and weeping. Originating from coastal plains north of Perth, the Illyarrie is extremely drought tolerant and handles the heavy clay soils of Grey's Plain well.

Zebra Finches are common in almost all regions of Australia. They make the most of good conditions following heavy rainfall and are found in large flocks. They feed mostly on fallen seeds on the ground but may also take insects.

While Zebra Finches are common and widespread, these seed-eating birds have suffered declines, particularly in tropical and subtropical savannas and arid Australia. In Northern Australia, these declines have been correlated with grazing intensity.

Planting Eucalypts has attracted "cup-moths." The adults are brown and heavily-built, with marbled forewings. The stout, slug-like caterpillars are a bright green with a bright red head and tail. Each end of the caterpillar is armed with clusters of sharp spines.

When the caterpillar is disturbed, these spines are erected, but at rest, they are retracted inside tubes. The spines cause intense stinging if they touch the skin and may even cause large lumps to appear.

The cocoons are spun into a form closely resembling the gum nut of the tree on which it feeds.

Color Plate 1: The Solex project Flora and fauna.
(Photos provided by Graeme Chapman of naturalight@graemechapman.com.au)

The land is not a "far-cry" from its "waste land" condition, but it is no longer as inhospitable as it was in the early 1990s.

4.5   SUMMARY

This chapter has looked at the economic and environmental outcomes of the Solex project over the 10-year period, 2005–2015. It has found that, while solar energy harvest was much better than calculated according to scientific predictions for region, the harvest from wind resources was less than half of that anticipated.

In particular, solar energy harvest was improved by 25 per cent over the predicted performance due to the use of seasonally adjustable solar panel chassis. That design permitted exploitation of a southern azimuth of sunrises and sunsets in the lower latitudes. It suggests that the adjustable design could be adopted generally in tropical regions and investigations made as to its application in higher latitudes.

The chapter looked at the integration of wind energy systems with solar pv arrays using the same enabling equipment. It found that, while the "dual-fueled" system is feasible, solar harvest in Carnarvon is so much higher than comparative wind-generated energy that ultimately solar pv arrays are vastly preferred over wind. The wind trial was ultimately terminated, and the enabling equipment will be used to service solar pv arrays, which will replace the wind generators.

This chapter also examined the economic and environmental outcomes of "downstream processing" of the solar-generated electricity by using it to manufacture ice. The concept of downstream processing of primary products is found universally as wheat, livestock, and fruit and vegetables are processed into bread, meat, and processed foods.

The value added to the electricity by using it to manufacture a commonly used product was considerable. Ultimately, it turned an economically unviable concept into a commercially viable venture.

An additional benefit is that the Solex project is environmentally sustainable. It not only does not contribute to greenhouse gas emissions but significantly reduces them. Avoided economic and environmental costs are not directly beneficial to the Solex project but go some way toward benefitting the region generally.

The Solex project has experienced changes in political philosophies by the Western Australian government, which has had a direct effect on the economic viability of its electrical generation and sales. Those findings support the principles put forward in Chapter 2 under the discussion of factors affecting the TBL.

This chapter finds that it is not only the generation of renewable energy that is not economically viable but so too is its competitor—the fossil-fueled generation system operated by the State. However, the ability to generate "cheap" energy impacts not only on the reduction of greenhouse gas emissions but also improves the economic viability of industry as an alternative energy source. The ice-works demonstrates "an alternative use for alternative energy."

The following chapter looks at the impact the Solex project has had on the broader community as the community of Carnarvon, the northwest of Western Australia, and the State generally adopted solar pv systems. Solex pioneered solar pv grid-tied installations in Western Australia's Outback, the public and industry adopted it, and Australia is less dependent on fossil fuel than it was in 2002 when the idea was first considered.

What was once regarded as little more than a novel experiment, conducted by an eccentric person, into the world of renewable energy resources is becoming part of mainstream Australian society and industry.

CHAPTER 5—THE SOCIAL IMPACT: "THE FRUITLOOPS"

Genevieve Simpson and Lex Fullarton

"Since the individual's desire to dominate his environment is not a desirable trait in a society which every day grows more and more confining, the average man must take to daydreaming."

Gore Eugene Luther Vidal (Jr), US Novelist and Essayist (1925–2012)

5.1   INTRODUCTION

This section has been developed with Genevieve Simpson, a researcher from the University of Western Australia looking at perceptions of domestic solar energy, and in particular the ways community networks can be instrumental in driving uptake. It includes an analysis of interviews conducted in Carnarvon by Simpson with stakeholders and community members who installed solar pv systems. Quotes from community members, industry, and government are used to highlight points of interest.

This chapter looks at the influence the Solex project has had on the broader community's attitudes toward the substitution of renewable energy in place of fossil-fueled, and polluting, electricity generation.

The chapter starts with a look at the interactions between government, utilities, and incentive schemes that have influenced the adoption of domestic solar energy. It also examines the way the Solex project, and in particular its founder Lex Fullarton, influenced the rate of construction of solar pv installations in Carnarvon.

In doing so, it considers the way community attitudes changed from scepticism and disbelief, to enthusiasm and involvement in engaging in the construction of dispersed embedded solar pv installations within Horizon Power's Carnarvon distribution system. This section highlights the extent to which Fullarton, and the development of a community of "Fruitloop" solar installers, could be considered a model informal solar community organization by engaging the community in developing the "Fruitloop" project with individual community members receiving financial benefits.

It looks at how the community was made aware of how to construct their own independent solar pv systems and how Solex assisted them to install those system and meet the compliance standards required by federal, state, and local authorities.

The chapter considers the network of relationships, partnerships, and stakeholders engaged in the Solex project and in the process of Fullarton becoming a "solar champion." It also considers the formal recognition and media attention received by Solex and considers the influence Carnarvon may have had in the adoption of solar pv systems in neighboring communities.

5.2   POLITICAL BACKGROUND

In 2005, the Western Australian ALP government enthusiastically embraced and encouraged independent renewable energy installations. It provided significant fiscal and administrative support to the concept of combating air pollution and greenhouse gas emission. The then Minister for Energy, and later to become Premier of Western Australia, Alan Carpenter opened the Solex Carnarvon Solar Farm and encouraged the Western Power Corporation to "[focus] on the future and that focus includes a strong commitment to sustainable practices and to the development of renewable energy sources."[241] In office, Alan Carpenter was a strong supporter of renewable energy, and government agencies showed enthusiasm and cooperation in order to contribute to the uptake of dispersed embedded solar pv installations.

In 2008 a change of State government occurred, from the socially focused ALP, to the economically focused Liberal Party. In line with this change of government, policies affecting renewable energy deployment also changed. On election to office in 2008, the Liberal Party started winding back fiscal incentives for larger-scale renewable energy projects and set about dismantling vehicles of administrative support, including the Sustainable Energy Development Office, the Office of Climate Change, and renewable energy units in energy utilities.

The rollback of political support for privately owned, solar pv installations is an example of the prevailing Australian political philosophies in regards to the relative political viewpoints of the ALP and Liberal Parties as depicted in Figure 5. It exemplified the respective party political philosophies as to their views on the recognition of the existence of climate change generally and methods of combating greenhouse gas emissions specifically.

While the Liberal Party was elected to government in Western Australia in 2008, continued support for renewable energy was maintained by the Federal ALP government, which enabled funding for renewable energy projects via the electricity consumer-funded renewable energy target. The scheme required electricity retailers and large-scale consumers to submit certified renewable energy credits for a proportion of their electricity consumption.

A subcomponent of this scheme that supported domestic solar energy installations was the Solar Credits Multiplier. The Solar Credits Multiplier, introduced in 2009, provided an additional financial incentive for solar panel installations by multiplying the

---

[241]   Western Power, *Annual Report 2005* (2005) 31.

number of certificates these systems could create. For administrative convenience certificates were "deemed" to have been generated in advance according to installations' rated capacity and location in Australia. The number of projected annual certificates was extended over an expected period of production of 15 years.

Under the Solar Credits Multiplier, the number of eligible certificates was multiplied by a factor of up to five to increase the value of returns to the owner from the sale of RECs (STCs) as discussed in Chapter 2.

This multiplier was designed to reduce over time, and it ceased on January 1, 2013,[242] some six months earlier than originally intended. STCs could be created for installations up to 20 kW in regions not connected to "recognized" electricity grids.

A clear exception to the reduction in support for renewable energy prevailing under the State Liberal Party was the introduction of a premium net feed-in tariff for domestic pv users. The Liberal Party went to the 2008 election with a proposed feed-in tariff in response to the ALP committing to a similar scheme if elected and based on feed-in tariffs introduced in other States.

The feed-in tariff was initially delayed, however, based on public support for the scheme and that it was an election commitment a net feed-in tariff of an initial 40 cents/kWh was finally rolled out with applications open in July 2010.[243] The scheme was suspended just one year later in August 2011, with more than 65,000 homes connecting solar systems.[244]

The feed-in tariff was a financial incentive to promote installation of domestic solar systems but did not pay for the value of the electricity generated and fed into the grid. This "buyback" rate reflected the cost of generation in geographic locations and was paid for via the Renewable Energy Buyback Scheme, which varied between the two state-owned electricity retailers, Horizon Power and Western Power.

These various policies resulted in a sharp increase in the privately funded construction of residential and commercial solar pv installations in Carnarvon. By the end of 2009, some 21 installations with a cumulative generation capacity of over 230 kW had been installed in Carnarvon. By mid-2011, Carnarvon had installed roughly the equivalent of one of Horizon Power's "peaking plants."

---

[242] Australian Government: Clean Energy Regulator, *Creation of Small-Scale Technology Certificates* (2016) <http://www.cleanenergyregulator.gov.au/About/Accountability-and-reporting/administrativ e-reports/the-renewable-energy-target-2014-administrative-report/Creation-of-small-scale-technolog y-certificates> at January 13, 2016.

[243] Government of Western Australia: Media Statements, *Feed-in Tariff Scheme Provides Incentive* <https://www.mediastatements.wa.gov.au/Pages/Barnett/2010/05/Feed-in-tariff-scheme-provides-incentive.aspx> at January 16, 2016.

[244] Government of Western Australia: Media Statements, *Residential Feed-in Tariff Scheme Suspended After Reaching Its Quota* (2011) <https://www.mediastatements.wa.gov.au/Pages/Barnett/2011/08/R esidential-feed-in-tariff-scheme-suspended-after-reaching-its-quota.aspx> at January 16, 2016.

Despite the high level of support for renewable energy within rural and remote communities in Western Australia, and the Nation generally, Horizon Power introduced a moratorium to the connection of solar pv installations in 2011. Horizon Power cited that

> [I]mpacts on the distribution network due to PV systems [were] starting to emerge. [T]he concerns about these impacts were sufficient enough [for it] to apply a limit of 1.15MWp of distributed PV system capacity on the distribution network.[245]

With the limit in capacity reached at the time this statement was made, the moratorium was brought immediately into effect so that this change of policy effectively brought the expansion of solar pv installations in Carnarvon to a close.

In spite of the availability of RECs/STCs and improved financial returns for solar systems, the enthusiastic expansion of community-based embedded solar pv systems remains severely restricted for the customers of the regional State utility, Horizon Power. At the end of 2015, the moratorium remained in place as debate continued as to how to increase solar pv hosting capacity.[246]

5.3   PARTNERSHIPS FORMED AND STAKEHOLDERS ENGAGED

The success of the Solex project and the continued enthusiasm for solar development in Carnarvon would not have been possible without the sustained interaction between Fullarton and members of government and industry. More importantly, however, continued flow of information from Solex to stakeholders enabled better understanding of renewable energy, and in particular solar energy.

It has always been the intention of the Solex project to contribute to a better understanding of how grid-connected solar electricity interacts with incumbent electricity generation technology and network infrastructure, in the interest of advancing the use of renewable energy systems. It is for this reason that data collected by Solex were forwarded to and incorporated in the University of New South Wales Centre for Energy and Environmental Markets (CEEM) Study of Grid-connect Photovoltaic Systems—Benefits, Opportunities, Barriers, and Strategies final report.[247]

Furthermore, Horizon Power has assigned staff for direct liaison with Solex for problem resolution and consultation, with a staff member at Horizon Power noting that Fullarton provides him with information around expansion of his systems and the

---

[245]   Simon Lewis, "Carnarvon: A Case Study of Increasing Levels of PV Penetration in an Isolated Electricity Supply System" (Report prepared by The Centre for Energy and Environmental Markets, 2012), 4.

[246]   Regional Power Corporation trading as Horizon Power *Solar Energy* (2016) <http://horizonpower.com.au/help-support/solar-energy/why-is-there-a-limit-on-the-solar-capacity-for-each-town/> at January 13, 2016.

[247]   Robert Passey et al, *Study of Grid-connect Photovoltaic Systems: Benefits, Opportunities and Strategies* (2009).

state of the current grid. This staff member went on to note that Fullarton is always keen to share data and that it is interesting to hear what he has to say.

The level of contact between Solex and Horizon Power varies, with Solex undertaking daily liaison with general staff and periodical contact with senior management. The highest concentration of interactions was around the resolution of problems associated with the roll-out of embedded solar pv installations on behalf of community members. That reflects the role that Fullarton played as a "community translator" with respect to solar matters.

The Solex project played an important role in the dispelling of certain myths or preconceived notions of stakeholders involved. These examples are not intended to be derogatory to the individuals but rather illustrations of ingrained attitudes and beliefs. For example, Fullarton described one interaction where a visiting state utility manager observed the array in its "south-facing" position and commented: "You have your panels facing the wrong way." Fullarton's reply was, "I have them facing the sun, which way would you face them?"

On another occasion, a statement was reportedly made by a similar ranking official of a state utility that "the wind turbines are too close together." When asked why that was so, the reply was that turbulence from the leading turbine will disrupt the air flow to the ones behind. The observer was given the direction north and asked what alignment were they on. "Why east/west" was the note. "What is the direction of the prevailing wind, and what is the present wind direction?" "Southerly to southwest" came the answer. "In that case how does the wind passing over the sails of one turbine create turbulence for its neighbors? Do aircraft not have a number of engines placed side by side, they appear to function without disruption?"

Furthermore, an industry member from Perth noted the benefits of having someone from "outside the system" provide alternative perspectives on design options:

> I had to do all the [Clean Energy Council] courses and you start by doing everything by the text[book] and then someone like Lex comes along and says "well why don't you just do it like that." In terms of going into the nth degree of the system design, I think some of the things that Lex has taught me is that you can spend all your time doing that, or you can do this and put on two extra panels and you've saved yourself a day of engineering and worrying about it and you're producing more energy and it will cost you $500 and that's it, and I thought "well that makes a lot of sense."

These, and many other incidents experienced over the years of operation, indicated that ingrained beliefs in "indisputable facts" can be difficult to refute, even if they are clearly demonstrated to be in error.

In addition to providing information and engaging with those in the energy industry, Solex and Fullarton also sought to engage members of government and public sector workers. Interactions varied from promoting renewable energy projects in regional communities, to providing information to advocating on behalf of community members.

The Solex project was important in promoting acceptance of renewable energy by the local parliamentarian and further cementing his support for renewable energy on behalf of his constituents. This successful physical manifestation of the Solex project was integral to many community members accepting solar energy, with this parliamentarian no exception:

> He built it. You could actually physically see what he had built, and he was able to physically show how it worked. It's better than looking at paper. You can see it work. You can understand it. Time is limited, and people put things in front of you and you need to be able to just read the back of the book. It was like taking a lecture, with a professor standing right there. I should say Doctor.

Fullarton was also important in promoting the availability of the Gascoyne Regional Development Commission's renewable energy rebate. Fullarton's existing ties with the project officer assisted with ensuring that all community members were able to meet the requirements of the scheme, and Solex was also able to provide information of assistance to the Commission:

> He's got very good knowledge about that side of things which is always really good when you're trying to put up proposals and that kind of stuff and if you ever needed any facts or figures you could always go to Lex and he would be able to reel them off. That's really the involvement that I had had with him, just getting that technical information if I needed it for any stuff that I needed to provide our board or any reports or that sort of thing.

In addition to the role Solex has played in providing information to stakeholders, Fullarton was also a member of the Sustainable Energy Association, an organization to promote the interests of the renewable energy industry, and was familiar to its former CEO for being "quite active and vocal." It was through these various forums that Fullarton was further capable of promoting the interests of solar in regional areas.

The Solex project site continues to be available to those who wish to learn more about solar in Carnarvon. A record of all visitors is maintained and reflects the diversity of interested parties who have shown an interest in the project.

In addition to the influences that Solex has had with industry and the community, two particular relationships are examined in further detail below. They are the relationship with the local utility, Horizon Power, and a community self-help group—the "Fruitloops."

5.4 HORIZON POWER

Horizon Power became a separate entity from Western Power in April 2006. It immediately implemented differing policies on solar pv installation integration to the distribution grids from the much larger and more conservative Western Power Corporation, which had responsibility for electricity supplies to the capital city, Perth, and its environs.

Horizon Power took on the specific responsibility for rural and remote Western Australia. Its generation systems were not only remote but also tended to be smaller and independent of other power stations, with the exception of the North West Interconnected System (NWIS) servicing the mining communities of the Pilbara. Significantly, small remote town power stations tended to be diesel fueled.

Unlike the previous Western Power Corporation, which focused on wind and bio-energy technologies,[248] Horizon Power's management was quick to realize the value of privately owned embedded solar pv installations in displacing the volume of diesel consumed in those communities.

To indicate Horizon Power's policy toward the adoption and integration of renewable energy with its generation and distribution system, its 2006 Annual Report states:

> There are five wind farms operating in Horizon Power systems across the State and another planned for Coral Bay. Horizon Power is a major customer of the Ord Hydro power station, which supplies the company's needs for Wyndham and Kununurra. A private solar energy farm, commissioned in October 2005, has been contracted to supply electricity in Carnarvon.

> Horizon Power has structured its Power Purchase Agreements with independent power producers to allow for renewable energy to form a part of the generating capacity. Private investment in small domestic renewable energy systems is also being supported by the introduction of a buyback scheme.

> Horizon Power will support the identification and development of other renewable projects in its systems.[249]

The private solar energy farm is the Solex Carnarvon Solar Farm. Solex's modest abatement of 5 tonnes of greenhouse gas emissions for 2006 is recorded under the heading of Environment in that report.[250]

Horizon Power implemented a one-for-one "buyback" tariff that resulted in it paying the sale price of electricity to domestic customers for net energy exported to the distribution systems. It revised the upper limit for solar pv installations from less than

---

[248] Ibid, 15.
[249] Regional Power Corporation trading as Horizon Power, *Annual Report 2006*, (2006) 21.
[250] Western Power, above n 250, 15.

5 kW to 30 kW for grid-connected solar pv installations and began design and construction of its own solar farms.

> In 2011, Horizon Power completed construction on two power stations in the inland Pilbara towns of Marble Bar and Nullagine which combine solar technology with back up diesel generation. Reliability of power supplies in the towns have improved and are well within the reliability standards set by the independent regulator.
>
> The power stations at Marble Bar and Nullagine incorporate technology which converts energy provided by the sun. The technology being applied provides the highest solar penetration possible, with 65 per cent of the day time load to be met from solar energy.[251]

Those policies resulted in a sharp increase in the privately funded construction of residential and commercial solar pv installations in Carnarvon. By the end of 2009, some 21 installations with a cumulative generation capacity of over 230 kW had been installed in Carnarvon.

Ultimately, more than 80 installations with a total capacity exceeding 1 MW were embedded into the Carnarvon Town distribution grid before the moratorium was placed on further installations by the utility. The reason for that restriction was stated to be disturbances to the electricity quality and control caused by the "negative draw" and the intermittent nature of the solar-generated renewable energy. Opinions vary as to the significance of that concern. However, further discussion on that matter is outside of the scope of this book. The issue is briefly examined in Chapter 6 as a suggestion for further research.

5.5   THE CARNARVON "FRUITLOOPS"

Vince Catania, the Western Australian Member of Parliament for the region said:

> We have got an organisation here of people whose members are called "Fruitloops." Led by the Number One Fruitloop being Lex Fullarton, who has really taken it up to bureaucracy to show that renewable energy is the most efficient way to go, particularly in the regions. He's been able to debunk myths and go over hurdles that have been put in front of him and really educated the community on that resource that sits above us in the sky. Not only in trying to reduce power costs but also in ways you can expand the use of solar in other areas, such as producing ice.

Following the example set by the construction of the solar farm and solar pv systems installed by the Fullarton's, the community of Carnarvon have adopted embedded solar pv with a passion. The Solex project, and in particular its proponent Lex Fullarton, have been instrumental in promoting the adoption of domestic solar energy in Carnarvon.

---

[251]   Horizon Power, *Marble Bar and Nullagine Solar Power Stations*, (2016) <http://horizonpower.com.au/about-us/our-assets/marble-bar-and-nullagine-solar-power-stations/ > at January 8, 2016.

There were several key elements that contributed to the role of Fullarton and Solex in promoting this adoption, including the visibility of the plant, the investment contributed by Fullarton, the role Fullarton played within the community, the level of knowledge Fullarton had around solar energy, and the extent to which he was able to drive institutional support for solar, through the creation of an informal solar community and links with local agencies and stakeholders, as outlined in the previous sections.

Combined, these elements supported the concept of an informal solar community organization. Solar community organizations aim to reduce barriers to the adoption of domestic solar energy through two mechanisms: first, by acting as credible sources of information on the benefits of solar and mechanisms for its installation; and second, by campaigning to encourage its adoption.[252]

Community organizations promoting renewable energy have been assessed based on their ability to involve individual community members in the establishment, setting up, and running of a "project" (the "open and participatory" dimension) and the extent to which the benefits of a "project" will flow back to the community members (the "local and collective" dimension).[253]

The relationship between these two dimensions is shown in Figure 54 and demonstrates that the Carnarvon "Fruitloops" formed an ideal community renewable energy project by promoting both community involvement in controlling the uptake of solar and also by ensuring strong financial returns to householders.

---

[252] Daniel Noll, Colleen Dawes, and Varun Rai, "Solar Community Organizations and active peer effects in the adoption of residential PV" (2014) 67 Energy Policy 330–343.

[253] Gordon Walker and Patrick Devine-Wright, "Community renewable energy: What should it mean" (2008) 36 *Energy Policy* 497–500.

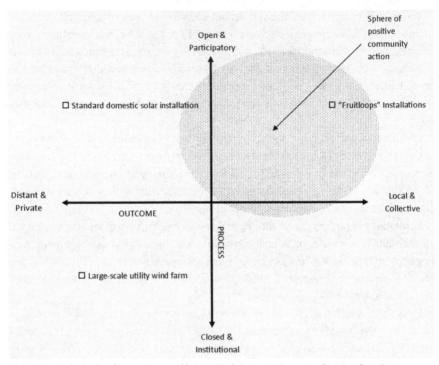

Figure 54: Understanding of community renewable energy in relation to project process and outcome dimensions. (Source: adapted from Walker and Devine-Wright[254])

The position of the Carnarvon "Fruitloops," depicted in Figure 54, is as an idealized community renewable energy organizational structure. It shows the extent to which the "Fruitloop" phenomenon was driven by community members, for the benefit of community members.

The social impact of the Solex project was almost immediate. Many of those who attended the official opening in August 2005 expressed interest in installing their own solar pv systems. The availability of the Solex ice-works was integral for community members in terms of building trust in solar technology, with one Carnarvon resident saying:

> I saw that the business was actually working, that it was making ice, we didn't go out there to see how it works, but we still think it's a great thing to have in Carnarvon, and [gave me confidence in solar].

254   Ibid 498.

The significant financial contribution made by Fullarton was also deemed as implicit confidence in solar technology by community members, with many confident of potential returns on their own systems based on Fullarton "putting his money where his mouth was." That feeling of financial risk assessment was echoed by one of the residents who stated:

> For Lex to put out a fairly large investment on something like that, it really piqued my interest.

Fullarton's evolving knowledge of the technical, financial, and administrative aspects of Australia's renewable energy schemes and regulations associated with deploying Solex were important contributors to the growth of domestic solar pv installations in Carnarvon. He completed training and registration as a Fronius Service Partner to enable him to service and repair Fronius equipment in Carnarvon. That provided a local service for the public without having to have their units serviced and repaired remotely from Melbourne in Victoria.

This benefit was perceived by the community, with some mentioning that:

> We stayed with Lex because if something went wrong we could whinge to him, because at least there would be someone to whinge to.

And:

> Oh, we all knew where he lived, so ... there would have been a picket line and torches if something went wrong!

He also received Clean Energy Council accreditation to design and supervise the installation of solar pv systems. This provides a local service without the community having to rely on installers visiting Carnarvon from other regions, including Perth. The financial benefit of having a locally skilled solar enthusiast, in addition to the local electrician, promoted prompt solar installations, with financial benefits that were positively perceived by the community:

> Oh yeah, it saved a lot of money I should imagine. Imagine how much money it would cost to pay for someone to come up from South, for all their accommodation and all that sort of stuff.

Solex organized and completed documentation for compliance with the design and installation of the roof-mounted installations with local, state, and federal authorities. Furthermore, Fullarton assisted with community members' receiving financial benefits by helping them with the completion of the various forms of paperwork for incentives and rebates and by finding buyers for their renewable energy certificates:

> Yeah, we didn't have to worry about the fine paperwork, and that sort of stuff, because Lex was covering all of that. He sorted all the paperwork out, and basically we didn't have to do anything, and that's why Lex was there, we paid him a fee for what he did and he got everything stamped and everything passed.

In particular, he achieved his Clean Energy Council registration that allowed him to trade RECs (LGCs and STCs). This provided a local service without the townspeople having to rely on agents to purchase and trade their RECs from other regions. It also permitted solar pv owners direct access to liable parties and permitted higher prices to be obtained for their RECs than those paid by agents who maintain a profit margin from buyers and sellers.

In this way, "the Fruitloops" relied on Fullarton's professional approach and business acumen. He had been a local taxation practitioner for nearly 20 years and was held in high regard by many in the community.

The introduction of the STC multiplier had the economic effect of reducing trade prices for them. For a period, the STCs traded as low as $15. This led community members to continuously engage with Fullarton to reach a best outcome for financial returns for their RECs. As one "Fruitloop" put it, prices for the STCs fell:

> But he kept letting us know what [the STC prices] were, they kept going up and down, you know what I mean. Some people got up to $30, and if you could sell them at the right time you could make a lot of money.

Furthermore, and considered most important by the community members, Fullarton provided a source of reliable and comprehensible information around the various schemes. This was reflected in comments such as:

> He has been the font of all knowledge and encouragement. Obviously other people have installed it and talked about it but Lex would be the expert.

And:

> Oh well, we did a bit of research, we were pretty well up with it, but if you go to someone who has already done those steps in front of you, you know, and a lot of other people [were] going to him as well because he had the connections to make it go, sort of thing. And we knew he had the connections. I wouldn't know how to go about it.

Most important for the community members in creating a sense of unification around the support for solar energy was the social aspect of installations, with homeowners carrying out the physical labor with the assistance of their neighbors.

Part of the arrangement was that Fullarton would help them on the condition that they helped each other. Community members described loading and unloading solar cells from trucks, measuring and drilling frames, and watching the electricians do their work:

> Because sometimes we were going out there to load them and he had a truck load of them and we were delivering them around town. We used to come up with pallets that were the right amount for each person, you know.

Qualified and licensed tradespeople carried out the electrical connections, with the co-operation of David Kearney of Carnarvon Electrics important in progressing the speed and number of solar installations in Carnarvon. Installations varied in design in accordance with roof characteristics but were almost all 5kW or multiples thereof, "split" arrays as illustrated in Figure 55.

Figure 55:   Typical 5- to 10-kW "split" solar arrays in Carnarvon. (Source: Photograph taken by Alexander R Fullarton, 2009)

Reducing labor costs by having householders help each other install systems, in combination with the maximizing of financial benefit of the RECs, where just two of the ways Fullarton helped improve the financial position of the community members when installing solar systems.

Materials were purchased in several parcels of 5-kW capacities to enable access to wholesalers, and panels and inverters were purchased directly from wholesalers by an arrangement established by Fullarton and shipped to Carnarvon as bulk freight. Installation costs were reduced to a minimum by creating scales of economy with up to four installations a day carried out.

Furthermore, Fullarton's strong voice and position in the local community influenced the development of an additional, region-specific renewable energy grant via the Gascoyne Regional Development Commission. Community members also took advantage of home extension loans and the Shire of Carnarvon decision to reduce building licence fees as an added incentive to make Carnarvon a "solar town."

The solar community of Carnarvon were initially dubbed "Fruitloops" for what was, at the time, considered to be eccentric behavior. However, once the economic and environmental benefits of independent renewable energy were realized by the broader community, others were quick to install their own system and adopt the badge of "Fruitloop":[255]

[255]   ABC News (Australia), "Carnarvon residents show solar power can pay," *7.30 Report*, April 13, 2013, <https://www.youtube.com/watch?v=1mS1D5yMjLg> at January 8, 2016.

When we first did it we were called Fruitloops because we were stupid for putting out money for the solar panels. Because people didn't understand solar power. And because it was so costly people were shying away from it. And then bagging we Fruitloops for doing it because we didn't know what we were doing. The proof of the pudding has come home!

5.6   RECOGNITION AND MEDIA COVERAGE

Solex, the Carnarvon "Fruitloops," and Lex Fullarton have received various awards and other forms of recognition for their contribution to renewable energy, energy efficiency, and community development.

Solex's achievements have been publicly recognized in the Western Australia Environment Awards 2008 as the winner of the Community Energy Efficiency Award. This Award acknowledges community groups that make a valuable contribution to protecting and conserving the State's natural environment, through reduced energy use.

Solex Carnarvon Solar Farm has also received two nominations, seven semi-finalist and one winner award for the Western Australian Regional Achiever of the Year between 2004 and 2008. The winning award was for the 2008 Regional Achiever of the Year Leadership and Innovation Award sponsored by Horizon Power.

In addition to awards directly attributed to the Solex project, the "Fruitloops" of Carnarvon also received public recognition as finalists in Western Australia's 2009 Regional Achiever of the Year Awards.

Fullarton has been active in consistently having Solex, the "Fruitloops," and the environmental and economic benefits of solar energy included in the media. The most significant exposure Solex and the "Fruitloops" achieved was in an almost eight-minute story on "7:30 Western Australia," the central current affairs program on Australia's public broadcasting station, the ABC. The story focused on the positive contribution solar power can make to reducing domestic power bills, the local community's support for solar energy, the role of Fullarton as a "solar champion," and the impact installed solar is having on Horizon Power's networks.

The role of Fullarton as "solar champion," as well as his key role as a community leader on the local council, was further highlighted in a 2015 article that designated Fullarton as a "Local Legend" of Carnarvon.

There has been further publicity throughout the region through local newspaper articles, which have had an important role in raising community awareness about solar energy and the Solex project:

I know Lex on a personal level as well, but I heard about his interest in solar through the media. He had been putting in articles and letters, and stuff like that, in the newspaper and then when Solex started there was always lots of positive stories about that.

Promotion is also made through radio and newspaper articles about associated re-newable energy projects in the area, with the ice-works used as a medium for demon-stration purposes. The Solex project has a commercial radio advertizement run on the local radio, which spans 600 km of the Northwest Coastal Highway.

5.7   INFLUENCE BEYOND THE "FRUITLOOPS"

The "Fruitloops" and their friends, neighbors, and colleagues in Carnarvon continue to show an interest in solar and invest where possible, with over 1,059 kW of solar pv capacity installed on sheds and businesses throughout the town of Carnarvon.[256]

Overall, by applying the formula for the displacement of fossil fuel and greenhouse gas emissions detailed in Chapter 4, the transition from fossil fuels to solar energy sees Carnarvon save nearly 490 tonnes of fossil fuel and over 980 tonnes of green-house gas emissions per year. This saving does not take into account the cost of transport and storage of the diesel fuel used in generators in Carnarvon.

The enthusiasm for adoption of domestic solar by the community has now spread to local institutions. The success of the trials and public acceptance has led the Shire of Carnarvon to plan its own installations with nearly 140 kWp of solar PV planned for buildings owned by the Shire. This includes its depot and the airport terminal build-ings.

Furthermore, schools in Carnarvon now install solar pv systems for teaching pur-poses and to displace some of the electricity they would otherwise purchase from Horizon Power. All of these activities are assisting Horizon Power to meet its renewa-ble energy commitments and to reduce reliance on fossil fuels and the production of atmospheric pollutants. An additional advantage is that this is being achieved from financial sources outside the State of Western Australia's revenue sources.

With the Solex project completed, Carnarvon has become a site of further solar experimentation. This includes a large-scale grid-connected solar system by Energy Made Clean. The project was proposed to be the "first large-scale power station in Western Australia to run entirely on solar energy" during its launch in 2010[257]. The plant was officially opened in May 2012, with 290 kW.[258] The project was completed well under its initial budget, reflecting the reducing cost of solar panels.

---

[256]   Solex data held by Alexander Fullarton.

[257]   Chalpat Sonti, "Solar power station a step closer to reality," *The Sydney Morning Herald*, February 16, 2010, <http://www.smh.com.au/environment/energy-smart/solar-power-station-a-step-closer-to-reality-20100216-o6in.html> at January 15, 2016.

[258]   Sean Smith, "EMC Solar to open Carnarvon Project," *The West Australian*, May 4, 2012, <https://au.news.yahoo.com/thewest/wa/a/13599005/emc-solar-to-open-carnarvon-project/> at January 19, 2016.

An employee of Energy Made Clean indicated that Fullarton and his engagement with the Solex project was integral to Carnarvon being selected as a site for the project, saying:

> It was really through him that the project was built in Carnarvon. There were probably actually some better cities for the Horizon network where it could have been more easily built but he was the reason why it was built there.

Fullarton gained a position on the board of Energy Made Clean during the development and commissioning of the Carnarvon solar farm and was deemed to be instrumental in engaging the local Horizon Power team to coordinate connection of the solar farm to the grid:

> It was always interesting to get his perspective on solar from where he is and the community he's amongst all the time, his perspective on Horizon Power and the politics around connecting renewable energy to the grid was always valuable. He gives us lots of feedback on what the opinion is out there.

Ultimately, the contribution of solar-generated electricity to the Carnarvon Town Distribution network was considerable. Figure 56 shows the combined output of the solar pv installations of Solex, "the Fruitloops," Energy Made Clean, the Shire of Carnarvon, and local schools in Carnarvon compared with total electricity generated and that generated by the state-owned energy utility.

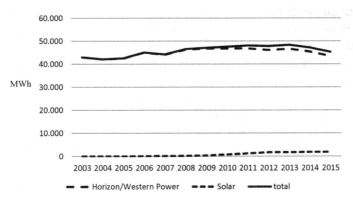

Figure 56: Annual fossil-fuel and renewable energy electricity generation for Carnarvon: 2003–15. (Sources: Western Power[259]; Horizon Power[260]; and Solex data)

---

259 Western Power, *Reports and Publications: Annual Reports* (2016) <http://www.westernpower.co m.au/corporate-information-annual-reports.html> and <http://www.parliament.wa.gov.au/publicatio ns/tabledpapers.nsf/displaypaper/3621468a5f3bae3bbab5a64f48256d9c00323e96/$file/west-ern+power+corporation+ar+2003.pdfat> at January 2, 2016.

260 Regional Power Corporation trading as Horizon Power, *Reports and Publications: Annual Reports* (2016) <http://horizonpower.com.au/about-us/overview/reports-publications/annual-reports/> at January 2, 2016.

The potential influence the Solex project and the "Fruitloops" of Carnarvon may have had on surrounding regions is difficult to ascertain but likely to have contributed to some level of increased awareness and enthusiasm for solar.

It is noted that, while Energy Made Clean indicated, at the time of opening its solar farm in Carnarvon, that it would like to develop complementary technology in Carnarvon, it is a Chinese company, Lishen Power, which is set to continue the solar experimentation trend in Carnarvon.

Lishen Power has a plan to install giant batteries to store electricity produced by solar systems that is surplus to requirements.[261] The plan consists of the installation of two battery units to be housed in shipping containers. These will potentially offset the need to run the current combined gas/diesel generator and act as "smoothing" devices to counter the variances inherent in renewable energy generation systems.

Fullarton traveled to Broome to present a case study on his pioneering solar farming in the Gascoyne to support the "Broome Community Solar Forum," run by Low Carbon Kimberley, with the intention of having "renewable energy experts propose a vision for Broome as the next town in Western Australia to embrace solar power and join the phenomenon of community owned power generation."[262] He has also presented to geography students of the Edith Cowan University on field studies in Carnarvon.

Efforts by Fullarton to inform other regional communities about the success of the Solex project and potential benefits of solar power are likely to have influenced adoption rates. Dispersed embedded solar pv installations are now not only on residences but are on market gardens, retail stores, and light industry. Additionally, the concept of solar energy for ice manufacture has been adopted in Broome and Exmouth.

Like Carnarvon, the enthusiasm for solar in a number of these towns resulted in Horizon Power having concerns about grid stability and putting in place a moratorium on new grid-connected solar systems, including the towns of Broome, Exmouth, and Denham. One northwest town after another quickly reached it "hosting capacity," and the roll-out of solar pv installations came to an abrupt halt.[263]

Figure 57 illustrates the cumulative uptake of solar pv installations in the postcode areas of Broome, Carnarvon, Exmouth, and Denham. The moratoriums placed on the

---

[261] Daniel Mercer, "Sun, giant batteries to power Carnarvon," *The West Australian*, November 24, 2015, <https://au.news.yahoo.com/thewest/wa/a/30178788/giant-batteries-to-store-power-geraldton-in-trial/> at January 19, 2016.

[262] Nicola Kalmar, "Solar project gets boost from local business," *The West Australian*, <https://au.news.yahoo.com/thewest/wa/a/19274310/solar-project-gets-boost-from-local-business/> at October 6, 2013.

[263] ABC News (Australia), "Carnarvon residents show solar power can pay," *7.30 Report*, April 13, 2013, <https://www.youtube.com/watch?v=1mS1D5yMjLg> at January 8, 2016.

installation of grid-connected pv systems are evident at the points where the number of systems no longer increases.

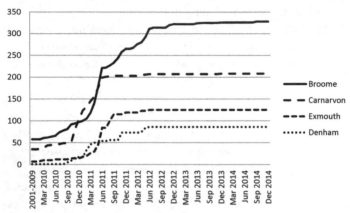

The Australian community has continued to show significant support for renewable energy and in particular solar energy, with solar pv installations continuing to roll out across Western Australia. It is suggested that the visibility of the Solex project, and other projects in the Carnarvon area, appears to continue to have an effect on the level of acceptance of solar energy in the community, as well as in surrounding communities. That influence was evidenced by a Carnarvon community member who said:

> A lot of people who come in where I work, they say "oh, you've got a solar farm" and it's because you can see it from the road coming and you can see it from the air, and they say "oh, you've got a solar farm, do you know who owns it can we have a look?" So tourists are sort of wanting to go and have a look.

5.8 SUMMARY

This chapter has examined the social interactions between the Solex project, its proponent Lex Fullarton, and various stakeholders.

In particular, this chapter detailed changes to incentive schemes for the support of renewable energy in Western Australia, highlighting that incentives helped drive significant uptake of domestic solar pv in particular, given the number of schemes available.

More important for the adoption of solar pv in Carnarvon, however, was the role Solex and Fullarton played in promoting the benefits of domestic solar pv and facilitating its installation. Solex/Fullarton provided administrative support via assistance in completing forms for the installation of systems and by accessing highest-value returns on STCs. Furthermore, Fullarton provided reliable information to community members and facilitated the provision of wholesale prices for solar components. In turn, Fullarton required that community members assist each other with the installation of their solar systems, which led to the development of a network of "Fruitloop" community members who were solar enthusiasts.

This chapter also detailed Fullarton's positive influence on the perception of solar energy and the transfer of knowledge to government and industry networks. Fullarton had an ongoing engagement with the local energy utility operators to share information and ease the connection of solar systems. He was also involved with communicating data to national research bodies and in advising industry members focused on both large-scale and small-scale pv systems.

The work of Solex, Fullarton, and the "Fruitloops" has been recognized with various awards, commendation, and through the media. Furthermore, it has resulted in significant increases in the adoption of pv in Carnarvon. The further expansion of pv in Carnarvon has been limited by the initiation of a moratorium on grid-feed-in pv systems in Carnarvon and surrounding regional communities. There continues to be work underway to resolve grid-stability issues and allow further solar penetration on these grids, including the potential for addressing issues via a large-scale battery system to be trialled in Carnarvon.

Solar pv adoption continues across Australia. In January 2016, Curtin University announced that the cumulative rooftop solar installation across the State is now Western Australia's "biggest power station," with 1750 new domestic solar systems installed per month in 2015, and some form of solar sitting on approximately 20 per cent of all homes across the state.[264] It is fair to assume that the leadership displayed by the Solex project may have played no small part in that acceptance and demonstration of success.

---

[264] Ian Clover, "Western Australia's rooftop solar now state's 'biggest power station'" *Renew Economy* <http://reneweconomy.com.au/2016> at January 7, 2016.

CHAPTER 6—REVIEW, RESEARCH CONTRIBUTION,
AND SUGGESTED AREAS
FOR FURTHER RESEARCH

"I must begin with a good body of facts, and not from a principle (in which I always suspect some fallacy) and then as much deduction as you please."

Charles Robert Darwin, British naturalist (1809–1882)

6.1   INTRODUCTION

By 2015, the Solex project had far surpassed its original goals and was well on the way to demonstrate that traditional fossil-fueled industry could not compete with a solar energy-based industry. Not only did it survive in a "damaged" physical environment but was able to restore much of the native flora and fauna to levels prior to those of European colonization of Western Australia in the coastal town of Carnarvon.

A sustainable enterprise is one that can continue indefinitely, not limited by the exhaustion of inputs or the creation of a degraded environment. The Solex project reduces not only the reliance on fossil fuels but also the impact of water consumption, pollution, and the handling costs of storage and transport of fossil fuels. It also reduces the environmental imprint of other organizations, chiefly the energy generation utility.

Furthermore, it created a social benefit by way of creating employment for one full time person, which had previously not existed in the community. The Solex project is a demonstrated business model that ties the triple bottom line (TBL) together.

6.2   REVIEW

Chapter 1 introduced the concept of the *TBL* and how each group of influencing factors are represented by the various Australian political organizations. It introduced the concept of how the "Socialist Left" ALP, the "Industrialist Right" Liberal Party, and the "Environmentalist Greens Party" viewed and supported or opposed renewable energy. It showed how the community of Carnarvon embraced the concept of dispersed embedded solar pv installations and placed the rate of construction within the broader scope of the Western Australian and the Australian population.

It then outlined the structure of the book and the discussion of Australia's renewable energy target in accordance with Australia's attempt to redress the problem of greenhouse gas emissions caused by the burning of fossil fuel.

In **Chapter 2,** the elements of the *TBL* and how the factors relate to each other were examined in detail. That discussion involved a history of the development of the TBL and how the respective political interests and philosophies have influenced the enactment of enabling legislation to combat the concept of global warming caused by greenhouse gas emissions.

It was found that, in Australia, rather than government acting as a cohesive force to move toward a coherent sustainability development framework, fragmented political interests tended to destabilize the harmony of the TBL.

Chapter 2 examined the carbon cycle and how greenhouse gas emissions have contributed to global warming. It looked at methods aimed at reducing existing CO_2 levels and how alternative energy usage and electricity generation methods can assist in that goal. Specifically, it looked at Australia's renewable energy target and the trading of carbon credits to reduce the impact of greenhouse gas emissions caused by Australia's reliance on fossil fuel-based electricity generators.

It examined the operation of the RET from a perspective of the implementation of the carbon tax and subsidy system established by the *Renewable Energy (Electricity) Act* 2000 (Cth) and the sales and marketing of carbon credits. It looked at the methods to account for the revenue earned from the creation and sale of those credits by the Solex project and the impact of the "tax system" on energy providers and consumers.

Finally, it established a tax-effective value of carbon credits created by the operations of the *Renewable Energy (Electricity) Act* 2000 (Cth) and the value of those credits to renewable energy generators by using the case of the Solex project and the financial data gathered over 10 years of operations.

Chapter 3 described the geographic environment of the Solex project by looking at its physical landscape and climate. It found that climate data recorded by the Australian Bureau of Meteorology for Carnarvon in Western Australia over a period of 62 years indicated an average trend rise in the annual mean maximum temperature for the location to be around 1°C. That rise in mean temperature is consistent with generally accepted climate change forecasts.

Chapter 3 described the history of the land on which the Solex project is situated and how the project was conceived and developed. It also looked at some of the technical aspects of how the solar farm was designed and constructed. The chapter was structured to outline the construction stages over the 10-year period from 2005 to 2015 including innovations introduced to the solar farm. Innovations included integrating wind turbines into the solar pv enabling equipment and "downstream" product enhancement by diverting solar energy into manufacturing ice to improve the revenue stream of the Solex project.

The chapter also provided estimates of expected production calculated by applying the scientific principles, which have been adopted by government agencies for the purpose of forecasting solar energy harvest in the region.

Chapter 4 presented the economic and environmental outcomes of the Solex project. It compared the production of the stages of development described in Chapter 3 with actual data collated over the 10-year period under review. It reveals the higher than expected performance of the solar pv systems and the less than expected performance of the wind turbines integrated into the solar farm.

It revealed how wind energy systems can be integrated into solar pv installations to extend the period of renewable energy beyond daylight hours. It also looked at the increased economic benefit of using "raw" electricity to manufacture products "on farm," thereby increasing the overall economic benefit of the Solex project. It was found that, for Carnarvon, solar pv installations vastly outperform wind energy harvest.

It also looked at the environmental benefits of solar energy production by way of displacement of fossil fuel consumption and related avoidance of greenhouse gas emissions. It considered other environmental considerations such as revegetation of degraded land using existing rainfall conditions. Finally, it compared the actual economic viability of the Solex project with the projections prepared by government agencies at the beginning of the planning stages of the project in 2003.

In particular, it found that the innovation of adding an ice-manufacturing facility changed the economic dynamics of the solar farm considerably. The innovation changed the prime directive of the project from producing "raw electricity" for sale, to the production of a manufactured commodity using the "raw electricity" as an input to manufacture rather than being a saleable commodity per se. The Solex project was changed from what was once considered extremely unviable to an economically sound business model.

Chapter 5 was co-authored by Simpson who studied the social impact of Fullarton and the Solex project on the broader community of Carnarvon and other towns in the Northwest of Western Australia. Simpson conducted her research by way of a series of interviews with various stakeholders in the community, industry, and government agencies.

She found that the Solex project played a pivotal important role in promoting the benefits of dispersed embedded solar pv and facilitated its installation. The chapter also detailed Fullarton's influence on the perception of solar energy and the transfer of knowledge to government and industry networks.

She suggests that, as solar pv installations continue to roll out across Australia, it is fair to assume that the leadership displayed by the Solex project played an important part in initiating public acceptance of solar energy and a demonstration of its success.

6.3   CONTRIBUTION TO RENEWABLE ENERGY SYSTEM DEVELOPMENT

This book provides a contribution to the understanding of the move away from centralized fossil-fueled electricity generation systems toward dispersed solar pv installations.

First, the case study provides a general insight into the conception and development of the first privately owned, commercial solar farm in Australia. The book outlines the problems faced and how they were overcome to provide guidance for others to replicate similar projects. It addresses the issues from a perspective of compliance with the sustainable development framework known as the *TBL*.

Second, it examines the concept of the TBL and introduces a fourth, overarching factor that forms the TBL into a coherent structure, which harmonizes the relationship between the economic, social, and environmental factors, which make up the TBL—the political environment. As illustrated in Figure 5 political influences can coordinate and harmonize the separate factors of the TBL to tie the sustainability framework together as one unit.

Third, it considers Australia's renewable energy target legislation and examines it from the perspective of a taxing and government expenditure structure. The book details how the market structure of creating, trading, and eventual surrender of renewable energy credits works and suggests how the market is reported in the revenue accounts of renewable energy generators.

It illustrates the operation of the legislation in monetary terms, to provide an insight into the renewable energy certificate market functions, and reveals the impact of the influence of changing political philosophies on trading prices.

Fourth, by illustrating various solar pv framework designs, it provides alternatives to fixed energy solar pv structures to provide methods to maximize energy harvest in varying seasonal conditions.

Fifth, and most importantly, it demonstrates an application of solar pv systems not to merely displace the use of fossil-fueled energy, and its related greenhouse gas emissions, but to exploit the renewable energy resource energy to create new industries.

6.4   RESEARCH LIMITATIONS

The scope of research of this book was limited to a case study of the development of a solar farm in Carnarvon Western Australia in 2005. It has looked at how the project was initiated, developed, and diversified to value add to its primary product—electricity to produce a marketable consumer product—ice.

It has followed the progress of the project over a 10-year period from 2005 to the end of 2015. It detailed some of the obstacles addressed in the development and operation from those of the environment—floods and tempests to administrative and financial compliance matters placed on it by economic considerations and bureaucratic philosophies.

It also looked at some of the broader social influences and impacts that the development had on the community of Carnarvon and the northwest of Western Australia. It had a direct involvement in the uptake of community owned solar pv installations in Carnarvon, as well as assisting with administrative compliance with obligations imposed under the *Renewable Energy Act (2000)* (Cth) for the creation and sale of carbon credits.

The book also examined the operations of the *Renewable Energy Act (2000)* (Cth) for the purposes of analyzing Australia's carbon tax legislation. However, that analysis was restricted as to how the legislation impacted on the Solex project and the owners of small-scale solar pv systems. It did not consider the financial or compliance costs imposed on generators of fossil-fueled electricity, defined as "liable parties."

Research as to the broader impacts on utility distribution grids was also limited to the data collated in Carnarvon. It is acknowledged that those findings cannot be applied in a general sense for large interconnected networks or smaller systems than the 16-kW State fossil fuel-powered station isolated in Carnarvon. Rather, it recognizes that limitation and recommends further investigations be conducted.

Solex successfully trialled a novel solar panel array structure, which was developed to take advantage of the ability to be adjusted to a south-facing azimuth during the summer solstice. While this system proved extremely successful at Carnarvon, further research will be necessary to establish the value of that system at latitudes higher than 25°.

Generally, this book represents a singular case study and acknowledges further quantifiable research is required to establish reliable scientific data to support or refute its findings for the development of solar pv renewable energy harvesting in a broader context.

6.5   SUGGESTIONS FOR FURTHER RESEARCH

The following areas of research are suggested as topics considered in this book but have been beyond the scope of this case study. They are vital to progressing the concept of sustainable development in Australia and the world and should be investigated with further research.

Desertification

While the concept of desertification, livestock grazing, and climate was briefly discussed in Chapter 3, to provide background and context for the development of the Solex project, those environmental factors of desertification were beyond the scope of this study.

The chapter considered the findings of Sturman and Tapper,[265] Savory, who considered that desertification can occur despite high rainfall events,[266] and Mitchell and Wilcox, who stated "growth not consumed [by livestock] is not wasted, but is actually necessary for the maintenance of good range condition."[267]

The findings and perspectives of those researchers are found in the following observation made by John Craig, a pastoralist, at Marron Station in the West Gascoyne, who said:

> A rainfall event has a greater environmental effect upon the rangeland than stock. Rainfall, or the lack of it, is more pervading. Livestock are selective and limited in their consumption of food on offer (due to the restraint of domesticated animals by fencing, and the restricted height to which they can graze). Furthermore, if vegetation becomes too sparse, they die. Nature becomes self-limiting.

> Over time, it appears a cycle of around 30 years becomes apparent. Consequently, the events of the 1930's might have similarities to the 1960's and 1990's. A pattern emerges of "unseasonal" heavy summer rains producing exceptional grass growth, followed by increased fuel loads, lightning from summer thunderstorms then ignites extensive bushfires.

> I have observed that summer rains produce grass; and winter rains produce herbage. A hierarchy of plant species develops, from short, prickly unpalatable species, to tall timber trees that shade out the ground cover in the dry environment. Many species have developed methods to protect their root systems from the intense ground heat of summer. One method is to de-foliate, as they become more water stressed. Shrubs "shut down" or become dormant so that they display little sign of life.

> Nature survives because of an innate survival mechanism of "balance". Trees die, trunks rot, branches fall to cover and protect juvenile seedlings. Flooding rains sweep up seedpods across the flats into a row of litter that humidifies and germinates a species. Cockatoos ringbark branches of trees; cyclones and strong winds break of the branches; white ants devour the centres to form hollow logs that then become nesting places for the next generation of birds.

---

[265]   Sturman and Tapper, above n 101.
[266]   Savory and Butterfield, above n 105.
[267]   Mitchell and Wilcox, above n 99, 19.

So whilst stock may contribute to "desertification" in some way by consuming limited vegetation for their sustenance, of more significance over time would be changes to rainfall. If rainfall events become related to a temperature variation, then the entire landscape undergoes a massive reorientation (which is beyond the scope of this dissertation to explain!).

The Gascoyne region has historically been viewed as one of great variability because of the inter-relationship between winter's frontal rains from the southwest, and the tropical north-west cloud bands and cyclones forming during summer.[268]

Craig is a long-term resident of the Gascoyne Region. He operates a pastoral property around 100 km from Carnarvon, which has been in the family for generations. His observations cover a span of some 65 years as well as anecdotal evidence passed to him by his parents.

These observations provide an interesting insight into the cyclical effects of climate and the impact of livestock on desertification.

It is suggested that further research could be conducted into the relationship between rising climate temperatures, rainfall, and desertification in the Outback of Australia, to investigate the relationship between the combustion of fossil fuels, greenhouse gas emissions, and the impacts of climate change on the arid regions of Australia.

Electric Vehicles

The concept of electric vehicles is considered briefly as a practical application of renewable energy. Further investigation could be made into the construction of "solar-powered" charging stations. It is envisaged that this would provide another avenue for an "alternative use for alternative energy and a future direction in solar pv energy to 'refuel'" electric motor vehicles.

Mortimore states that:

[t]he transport sector is arguably the most difficult and expensive sector in which to reduce greenhouse gas emissions (GHG), with carbon dioxide (CO_2) generated by transport in Australia increasing by 50.7 percent (93.5 Mt CO2-e) in 2012–2013 from 1990 levels (62.0 Mt CO2-e). Unless the government reverses this trend, CO2 emissions will continue to rise and offset the gains made in reducing carbon emissions in other energy sectors. The largest contributor to transport GHG emissions is road transport.[269]

She suggests "that vehicle taxes [could be] reformed into an environmental tax." She states that a tax based on vehicle greenhouse gas emissions, "is a "powerful instrument" that can "drive consumer demand towards fuel efficient cars" and foster a more sustainable car market."[270]

---

[268] Email from John Charles Craig to Alexander Robert Fullarton, January 13, 2016.
[269] Anna Mortimore, "Reforming vehicle taxes on new car purchases can reduce road transport emissions—ex post evidence" (2014) 29 *Australian Tax Forum* 177, 179.
[270] Ibid, 216.

The system of using punitive taxing system to bias market forces by way of artificially increasing prices may have some merit. However, price elasticity or consumer demand sensitivity to prices can be problematic. The history of tobacco and alcohol taxes and their impact on consumption indicates that consumers may not be as responsive to price increases on luxury goods as may be envisaged. Regardless the success of that desired economic measure, Mortimore's suggestion is acknowledged.

However, this book observes that, while electric cars are a step forward in reducing greenhouse gas emissions created by motor vehicles, the electricity to refuel these electric cars is derived from fossil fuel generators. The technological "leap forward" to a pollution-free future for motor vehicles will only arrive when they are "refueled" from renewable energy sources.

To illustrate that point, on the 26th of September 2014, one of the support cars of the Australasian Safari 2014 car rally was "refueled" with solar generated renewable energy at the Solex Carnarvon Solar Farm. The makeshift charging station depicted in Figure 58 could be industrialized and commercialized along similar lines to the existing "fossil fuel" gasoline pumps at roadside fuel stations.

Figure 58:   "Refueling" the first electric vehicle to arrive at the Solex project September 26, 2014. (Source: Photograph courtesy of Mitsubishi Motors Australia Facebook page[271])

---

271   Mitsubishi Motors Australia Facebook Page (2014) Facebook <https://www.facebook.com/Mitsubishi MotorsAustralia/?hc_location=ufi> at January 8, 2016.

The 2014 Mitsubishi Outlander petrol hybrid electric vehicle (PHEV) has an extremely efficient fuel consumption ratio at less than 2 L per 100 km. That compares very favorably to the average saloon or sport utility vehicle (SUV) at 6–8 L per 100 kM, or an average four wheel drive vehicle at over 10 L per 100 km. However, it relies on either an on-board gasoline-fueled generator or a fossil-fueled central power station for its electrical charge.

To be truly environmentally sustainable, electric cars can only be efficiently recharged with electricity sourced from renewable energy sources, to eliminate greenhouse gas emissions from road transport.

Unfortunately for Carnarvon, it has been reported that

> [t]he mayor of the NSW city of Dubbo has laid claim to being the first in Australia to drive a fully electric mayoral car after his council invested in a Nissan Leaf just before Christmas.

> The $27,999 LEAF is the third electric vehicle for Dubbo Mayor Mathew Dickerson—following on from a Mitsubishi Outlander, and before that a Holden Volt—but the first to be 100 per cent electric, with no back-up petrol engine.[272]

This has prevented another "first" for solar energy in the region. Given the vast distance between Carnarvon in Western Australia and the City of Dubbo in New South Wales, it is unlikely that the Solex project directly influenced the Dubbo city council in its decision to acquire the vehicles. However, but the combined examples of the Solex project, in Western Australia, and the City of Dubbo, in the eastern state of New South Wales, should have a strong influence in the transition from fossil-fueled motor vehicles to electrically powered transport.

As stated earlier, the transition to electrically powered vehicles only displaces fossil fuel when renewable energy sources are used to recharge the vehicles. In the case of the vehicle owned by the City of Dubbo, referred to in the article, it is primarily charged from a domestic solar installation.

Electricity Grid Stabilization

Chapter 5 outlined the basis on which the utility—Horizon Power—placed a moratorium on further solar pv installations being connected to its distribution system. Key to the cessation of further grid connected solar pv systems, and part of the reason for reducing tariffs for the purchase of solar pv generated electricity, is that it may destabilize the grid.

---

[272] Sophie Vorrath, "Dubbo mayor drives fully electric Nissan Leaf, powered by the sun," *Renew Economy* <http://reneweconomy.com.au/2016> at January 25, 2016.

Horizon Power's rationale is that

[w]hen a cloud covers the sun, the houses with solar panels may reduce the amount of electricity they are generating.

This can place a very sharp demand on the power station which then needs to quickly compensate for the sudden loss in electricity generated by the solar panels. Any engines not operating will need to start generating power again. However, this takes time and the power station may be unable to generate enough electricity to meet the demand on the network.

In this situation, the reliability and security of power supplies to all customers may be affected and customers may experience an unplanned power interruption.[273]

Horizon Power points to a limit of "hosting capacity," which is considered to be a maximum capacity of "unmanaged" solar pv installations the fossil-fueled generation system can tolerate to ensure disruptions do not occur due to the movement of clouds.

The discussion in Chapter 4 as to the integration of wind energy generation into the solar pv systems acknowledges variations in energy flows, in particular the vagrancies of wind energy harvest. Figures 44 and 46 point to the relatively smooth parabolic curves of solar pv systems in Carnarvon and refute the principle proposed by Horizon Power as to instability.

However, clouding problems do occur, and while Figures 44 and 46 do not reveal such variations, there are a number of days when intermittent cloud can result in variations similar to that shown in Figure 45.

The concept of intermittent generation and grid disturbances was considered in a report produced in 2007, by the University of New South Wales for the government of Western Australia.[274] The report pointed to "important issues of islanding prevention, timing and stability which are emerging as issues to be further investigated as the number of grid connected PV systems increases in the distribution system."[275]

However the report also noted that

PV's contribution to power quality (both positive and negative) is largely determined by the type of inverter used to connect to the grid. If inverters that conform to Australian Standards are used, PV is unlikely to have a significant negative power quality impact and should not result in inactive networks becoming "live" and therefore dangerous. PV can have a positive power quality impact only when it is operating. It may not provide the type of support required and there are generally cheaper alternatives available. Thus, while PV may be able to help

273   Regional Power Corporation trading as Horizon Power, *Understanding Renewable Energy* (2016) <http://horizonpower.com.au/being-energy-efficient/solar/understanding-renewable-energy/> at January 15, 2016.
274   Robert Passey et al, *Study of Grid-connect Photovoltaic Systems—Benefits, Opportunities, Barriers and Strategies: Final Report* (2007) 3.
275   Ibid, 106.

with power quality, this ability is likely to have little positive impact on its commercial viabil-ity.[276]

Futhermore:

> Reducing diesel use for electricity generation could have balance of trade benefits as most of the diesel used in Western Australia comes from the refinery at Kwinana and more than half the crude oil for that refinery comes from outside Australia. Increased deployment of PV sys-tems in rural and remote areas should result in local job creation, not only for installation but also for maintenance, and reduced electricity costs should also have economic benefits.[277]

The report notes that stability of the electrical grid was an emerging issue in 2007 and suggests that, and accompanying issues, be further investigated. It also found that solar pv had a positive contribution to power quality and fossil fuel-saving economic benefits. Those findings would appear not to support a restriction or limitation in the capacity of grid-connected solar pv systems.

In 2009, Passey et al stated that "PV is dependent on incident solar radiation and so is inherently intermittent. Thus, at high levels of penetration, PV could have an adverse impact on the reliability of electricity supply."[278]

However, they further state:

> While a high level of penetration is unlikely to occur at a system-wide level (and so significantly affect PV's value in offsetting conventional generation), it may occur in isolated grids (e.g. in rural areas serviced by diesel generation) and on particular feeders.

> The HOMER modelling software used here limits the degree of PV penetration by requiring that the PV/diesel hybrid system must be able to cope with an instantaneous 25% drop in PV output, with larger drops occurring gradually enough for the diesel generators to ramp up their output. The rapidity of such drops due to shading from clouds or aircraft can be reduced through distribution of the PV panels over as wide an area as practical.

They conclude that "[i]ncreased deployment of PV systems in rural and remote areas worldwide will have greenhouse gas reduction benefits and should also result in local job creation, not only for installation but also for maintenance, while reduced elec-tricity costs should also have economic benefits."[279]

Despite the caveat that high levels of pv penetration could have an adverse impact on electrical energy quality, Passey et al imply that the "high level" is around 25 per cent of the system capacity, and not the 5 per cent limit that Horizon Power appears to have implemented. The suggestion that aircraft shadows could cause sudden losses in pv energy generation also suggests very small solar farms indeed. This book also

---

[276] Ibid, 10.
[277] Ibid, 107.
[278] Robert Passey et al, above n 256, 427.
[279] Ibid, 430.

notes that it appears clouding results in a total loss of solar pv energy. That would only occur in the case of a total solar eclipse.

In 2011, Passey et al found that "Short-term intermittency of PV can be reduced through geographical dispersal. Very little or no correlation in output over 1min time intervals has been found for sites as little as 2 km apart."[280] Final they conclude that

> [t]here is increasing pressure to quickly implement DG on electricity networks, but to do this at medium to high penetration levels will require careful preparation and development of safe and carefully integrated protection and control coordination.[281]

In 2012, Lewis conducted another case study in Carnarvon (CEEM Report) as to the increasing levels of pv penetration in an isolated electricity supply system. He found that, in Carnarvon:

> [the] high PV penetration is coupled with a strong solar resource (an average daily solar inso-lation of 6.2kWh/m2). PV penetration is estimated to peak at 13% of system load at midday in both summer and winter. Consequently, impacts on the distribution network due to PV systems are starting to emerge. In 2011 the concerns about these impacts were sufficient enough for Horizon Power (the utility that owns and operates the Carnarvon distribution net-work) to apply a limit of 1.15MWp of distributed PV system capacity on the distribution net-work.[282]

Solex data indicate that the percentage of generation capacity of the Carnarvon town distribution network is around 1.06 MW of energy from the dispersed solar pv instal-lations and around 15 MW of deliverable fossil fuel generation from Horizon Power's 19-MW power station.

Therefore, unless the power station is not running at full capacity, which is the ra-tionale of having the dispersed solar pv installations then a maximum penetration of only 7 per cent is possible, about half of that suggested in the CEEM report 2012. An explanation for the apparent discrepancy is contained within the report. Lewis states "Carnarvon is an isolated network supplying approximately 5000 people, with an av-erage load of approximately 8MW in summer, 11 MW peak and 7MW in winter with a 4.5 MW low."[283]

It appears that the power station at Carnarvon, at that time, reached maximum demand at just less than 75 per cent of its deliverable capacity. It is not clear if the 1 MW of solar capacity has is included in those estimates.

The report also includes a key benefit to the community arising from the solar pv installations.

---

280 Robert Passey et al, "The potential impacts of grid-connected distributed generation and how to address them: A review of technical and non-technical factors" (2011) 39 *Energy Policy* 6280, 6284.
281 Ibid, 6290.
282 Lewis, above n 254, 4
283 Ibid, 40.

Historically the deration of the generating sets, particularly during February, has led to power outages in the town predominately due to the commercial bore fields sight connected to the grid. There is evidence from the community that suggests the frequency of these outages has decreased dramatically since the connection of 500kWp of solar in 2010 as this approximately matches the size of the bore field load. The reduction in the need to compensate for this large load by disconnecting customers has obvious benefits for customer satisfaction and network outage limits. By also adjusting other loads, such as irrigation pumping, peaks can be further shaved maximising the economic benefits for the utility.[284]

The report illustrates in Appendix 4 that the solar pv installations are well dispersed over and area of roughly 100 sq km. Clusters of installations focused in the town are generally of around the capacity of 5 kW, while those further out range up to 30 kW.

Twidell and Weir suggest that correlation of short-term variations for wind energy, a far more reactive energy source than solar as discussed in Chapter 4, is in the region of 20 kM separation to smooth out variations of up to 30 minutes.[285]

Boyle recognizes the variable output of solar pv systems and suggests that

> as long as the capacity of variable output power sources such as PV is fairly small in relation to the overall capacity of the grid (most studies suggest between 10 to 20%), there should not be a major problem in coping with fluctuating demand. The grid is, after all, designed to cope with massive fluctuations in *demand*, and similarly fluctuating sources of *supply* like pv can be considered equivalent to "negative loads." Such fluctuations would also, of course, be substantially smoothed out if PV power plants were situated in many different locations subject to widely varying solar radiation and weather patterns.[286]

In depth examination of this issue is beyond the scope of this book; however, as at the end on 2015, the matter appears to be unresolved in Carnarvon and other towns mention in Chapter 5. Of significance is a report that

> Immense popularity of residential solar in Western Australia essentially forms the state's largest power generating unit, according to research by Curtin University.

> Data published this week by Curtin University in Australia has revealed that the sheer volume of rooftop solar capacity installed in the state of Western Australia (WA) is such that, collectively, solar power comprises the state's de facto largest power station.

> Figures published by the Australian Clean Energy Regulator late last year show that Western Australia has more than 192,000 solar power systems installed, and research from Curtin University shows that there is 500 MW of PV capacity connected to the state's South West Interconnected System (SWIS).[287]

---

[284] Ibid,36. Despite being located in an arid region of Australia, as discussed in Chapter 3, Carnarvon has an extensive horticultural industry irrigated from underground aquifers situated along the predominantly dry, Gascoyne river. The bore field is powered by electricity the town distribution system.
[285] Twidell and Weir, above n 13, 304.
[286] Boyle, above n 108, 96.
[287] Clover, above n 273.

It is suggested that an independent investigation be conducted as to the impact of variability of solar pv installations to the grid, and an inexpensive, practical solution be implemented in the very near future.

Sustainable Architecture

Figure 8 illustrates a third method of reducing atmospheric CO_2 that of reducing energy consumption. Sustainable architecture to construct energy efficient buildings has not been addressed in this book, however, without some method of energy storage the diurnal nature of solar pv electricity generation limits energy access to daylight hours only. Solar energy does not "turn the lights on."

Considerable technological research is being applied to electrical storage by way of batteries or the application of other generation systems, including fossil fuel-based systems to "keep the lights on."

This book discusses the manufacture of ice as a possible application of energy storage beyond the hours of daylight. It has discussed the use of "daylight switches" to ensure the focus of energy sources away from fossil fuels to solar-sourced energy.

Ensuring energy use coincides with energy availability, or "load profile" shifting, requires more than energy storage to extend the influence of solar energy beyond daylight hours. One of the ways in which this can be achieved is by designing buildings to store energy.

By insolating buildings to reduce energy flows through roofs, walls, and floors, energy consumption is reduced. By increasing internal mass, energy can be stored as latent energy, which is absorbed or emitted to maintain internal ambient temperatures. Air-conditioning systems can be operated throughout daylight hours, when solar energy is available, to heat or cool internal mass so that energy can be absorbed to cool the building at night, or vice versa.

Buildings can be orientated in such fashion to maximize solar energy harvest from roof structures and to reduce or maximize energy absorption as seasons change. Research has been carried out on passive sustainable architecture; however, it is suggested that architecture consideration be made to the integration of active energy systems, such as air-conditioning systems energized by solar pv systems.

6.6   CONCLUSION

In the early 2000s, Fullarton acquired an abandoned piece of land on the outskirts of Carnarvon, which was once intended to be developed in the early days of colonization. This arid land was considered unfit for any "reasonable" purpose and thought to be "waste country."

However, after legally acquiring possession under the terms of adverse possession, Fullarton conceived and built a solar farm on the land to demonstrate the economic, social, and environmental benefits of renewable energy systems in developing alternative and positive uses for such "worthless" landscapes.

Not only did the Solex project ultimately prove to be economically viable but also socially and environmentally beneficial. From inception, the concept encouraged the broader community of Carnarvon to embrace energy independence by installing dispersed, embedded solar pv systems throughout the town and its environs.

The concept, at first supported by the State utility and government agencies, drew the attention of a small group of community members. Initially dubbed as "Fruitloops" for their acceptance of such an outlandish concept, the population of owners of solar pv systems flourished.

By March 2011, the State utility froze connection of further solar pv installations in Carnarvon. The northwest towns of Broome, Denham, and Exmouth quickly followed. In January 2016, it was reported that the total capacity of rooftop solar pv installations had become Western Australia's "biggest power station," with some form of solar sitting on approximately 20 per cent of all homes across the state.[288]

Simpson suggests (in Chapter 5), that it is fair to assume that the leadership displayed by the Solex project may have played a part in the acceptance of solar energy in Western Australia by its demonstrated of success.

---

[288] Ibid.

APPENDIX A

Carnarvon 15KW Solar Farm

"Your energy—free from the sun"

AWA Dish at Carnarvon reflecting the Town's space age history.

Proposal for the Construction of a 15kW Solar Farm at Lot 42 Boor Street Carnarvon Western Australia

by A R & J Fullarton

Revision 4: July 2004

Prepared by A R (Lex) Fullarton
Dip Real Est Mgt.; Dip Acct.; B Com.; PGrad Dip Com.; MTax.; M Com. FPNA FTIA AIMM
AREI CD
P O Box959
3 Crossland Street
Carnarvon WA. 6701

OVERVIEW: WHY A SOLAR FARM?

It is proposed to take advantage of Carnarvon's global position, climate and solar energy zone, to produce pollution free, environmentally responsible, electrical energy for use by the industry and citizens of the township and environs of the electrical distribution centre of Western Power's existing distribution infrastructure. A 15 Kilowatt solar power farm is to be constructed adjacent to Western Power's south eastern feeder line to supplement the utilities existing gas and diesel supplement power station.

The project is planned for long term fossil fuel free energy supplies which will deliver the highest supplies of energy onto Western Power's Carnarvon grid system at a time when the highest demands are being made on the system—that is the middle of the day. This will obviate Western Power's need to run and maintain extra fossil fuel engines at peak periods.

Solar energy development has been far advanced by the United States space program and now provides reliable, pollution free, electrical energy. Solar fans are used extensively in the United States, though adoption of solar energy conversion in Australia is limited. A pilot plant of 21kW production is installed at Kalbarri by Western Power. A number of solar plants up to 5kW have been installed in Western Australia under the Renewable Energy Buyback Scheme (REBS) though only two outside the Pilbara and South West distribution grids.

Initially the financial return to the proponents is negligible but the project is to be considered over twenty five years which is the guaranteed life span of the solar panels. As costs of energy escalate and reliance upon fossil fuels become less environmentally and socially acceptable, the profitability of the venture will soar. The major element being solar energy whilst requiring high initial capital outlay is fuel free, thereby proving to be highly attractive when viewed on a life-cycle basis.

> The first cost of solar devices is much higher than that of conventional fuel-operated devices. For example a small solar water pump, driven by a small reciprocating engine using vapour (sic) from a solar collector, in 1980 cost about $1,000 (US) per horsepower. An ordinary pump, driven by gasoline, engine cost between $50 and $100 per horsepower. However when "life cycle" costs are compared, the difference is not as great. Life-cycle cost involves capital cost plus the cost of energy and service supplied during the useful life of the device. Over 10 years the amount of gasoline used by a gasoline pump would amount to many times the initial cost of the pump; hence a solar water pump might prove to be less costly when life-cycle costs are compared. Where backup fuel-operated devices are used in addition to the solar equipment

(spinning reserve), the life-cycle costs of the solar device would be offset by fuel savings alone.[289]

This exciting project is not solely for the commercial benefit of the proponents but is viewed as a long term concept for the benefit of investors, Governemnt utilities and the community of Carnarvon.

Mr Fullarton is a descendant of Carnarvon's original founders and has a long and close association with the town. He is committed to environmentally favorable energy production as well as supporting and promoting industry within Carnarvon and its environs.

The technology has been lead by the United States space program and is well developed. The application of solar energy in the United States accounts for a significant contribution to annual usage but is yet to reach such levels of acceptance in Australia and particularly Western Australia. This solar farm will be by far the largest project completed in Western Australia and the first of its kind.

The following statement from the European Commission's report on solar energy production supports the view that Carnarvon should be exploited for solar energy.

Acceptable production costs of solar thermal electricity typically occur where radiation levels exceed 1700KW/m2 -yr. Appropriate regions include the southern European countries, North Africa, the Middle East, western India, Western Australia, the Andean plateaux, north- eastern Brazil, northern Mexico and the south-west United States. There are three systems of principal interest:

- The solar farm concept
- The solar power tower
- Parabolic dishes

A hybrid concept whereby a solar thermal power supply is integrated with conventional combined cycle fossil fuel generation is also being considered and is likely to be the most economically viable early scheme for the EU. This allows two stages of heating, which raises steam temperatures and hence operational efficiency and, most importantly, extends the operating hours of the plant.[290]

The solar panels are guaranteed against defects for a period of 25 years and are expected to render good service for many years after that.

Of important consideration is the basic concept that solar energy is rated at 1000 watts per hour per square meter of Earth's surface. All production figures contained

---

[289]   Macmillan Educational Company, *Collier's Encyclopaedia,* vol 21 (at July 4, 2004) Solar Energy, "Economic Considerations" [168].

[290]   *Solar Thermal Energy—Introduction* (1996) European Union, <http://europa.eu.int/comrnlenergy_tra nsport/atlas/htmlu/steint.html> at June 4, 2004. <http://helios.teiath.gr/creta/Library/html/ste int.html> at April 4, 2015.

within this proposal are based on that data, yet according to the ATLAS report Carnarvon should expect 1700kW/m /yr, a seventy per increase. That the solar farm proposed will produce 70% higher ratings will only be proven in actual operation. Such additional production is considered a "windfall" if it occurs. Information provided by the Sustainable Energy Development Office is that it is reasonable to expect an average of 6.5KW per day on a square meter of angled surface average over a year.

The figures produced in this proposal are slightly higher due to the low light intensity SQ 150 solar panels which produce a fixed 150 watts of power regardless of light intensity. The panels are engineered to produce no more than 150 watts of energy despite light conditions. Data extracted from a 1.92kW pilot plant by the proponents indicates that daily production averages 6.5kW in June to 10.5kW in December.

Not only will the project produce electrical energy at times critical to added demands, the middle of the day, but also produce "renewable energy credits" which can be used by Western Power to offset other conventional fossil fuel power stations in other locations.

This project is viewed as adopting 21st Century technology and should provide proof of Carnarvon's long standing leading role in advanced technology since it was pivotal as the NASA space tracking station involved in the Apollo moon landings in the 1960s and 1970s.

Carnarvon's electrical consumption rate, currently around 8MW, has been relatively static for many years as one project has evolved to take the place of another. The AWA communications station replaced the NASA tracking station, which was later replaced by Radio Australia's broadcasting station, in turn replaced by expansion of Dampier Salt's Lake Mc Leod salt operations and expansion of the fishing and fish processing industry. Dampier Salt proposes to construct a 1.5MW wind turbine power generation system which will greatly support Carnarvon's demand.

The intrinsic benefit of this structure will be to increase the profile of Carnarvon at a point on one of the major highways connecting the Pilbara and Kimberley to the South West. Easily viewed from the North West Coastal Highway, the project will inspire the interest of tourists. The site will be promoted accordingly by way of brochures and signage.

Corporate Structure

Initially the project will be that of a sole trader with a limited liability private company to be incorporated subject to the Companies (Western Australia) Code, as the project develops. The subscriber shareholders to be Master and Mrs A R and J Fullarton of 3 Crossland Street Carnarvon, by whatever interposed entity they see fit, in equal shares. The land will be leased to that company, on terms to be negotiated, by A R Fullarton for a period of not less than twenty five years, with an option for twenty five years at the expiration of that term.

2

The Site

The site consists of 40,473.3168m of flat clay pan, contained within a rectangular boundary being 256.29m true north and 157.92m east. The land is described as being Gascoyne Location 42 and is the whole of the land contained within Certificate of Title, Volume 664 Folio 74. The registered proprietor is Alexander Robert Fullarton of3 Crossland Street Carnarvon. The land is situated 305m south of Western Power's 22,000v and 240v service lines in Boor Street adjacent to the North West Coastal Highway. A lease has been granted by the Department of Land Administration for a twelve meter portion of land to be attached to Lot 42 and abutting Lot 34 and allows direct access to Boor Street along the Western boundary of Lot 34.

Apart from Lot 34 Boor Street, which is similar open scrub, there are no adjacent neighboring properties. The land form, that of river delta silt deposits, and lack of vegetation favors the development of the collection of solar energy unimpeded by shading by landforms, vegetation or manmade obstructions.

The global position of the site is approximately 24.880° South and 113.670° East, it is from this point that solar and meteorological observations and data from the Perth Observatory and the Bureau of Meteorology have been used to establish solar energy collection feasibility.

Production

Production of electrical energy will be by way of direct conversion of sunlight into electricity delivered directly to Western Powers' Carnarvon Town supply.

Western Powers' distribution system includes a 240v and a 22,000v power line which passes 305 m north of the site. Connection will be made by way of a suitable underground cable, metering and transformers to that distribution line. As outlined above delivery to the grid will reach its highest point in the middle of the day and will reach its longest delivery in the middle of summer, both at times of greatest demand.

A computerised Inverter/Transformer/Controller converts the Direct Current produced by the solar panels to Alternating Current in three phases, transforms the electricity to 240v and monitors, regulates and tracks the electricity supply of the distribution grid.

The proven technology automatically disconnects the solar farm from the grid in the event of power outages and a safety factor to workmen and others to prevent electricity flows in the event of intentional disconnection or accidental breakage of the grid.

It is intended to use **Shell SQ 150** solar panels for the conversion process. 'The panel operates silently without traditional fuel, waste or pollution. [It] is designed for use in a 24 volt system to perform efficiently in all operational conditions. Its distinctive ability to deliver . . . power levels in low-light situations makes it particularly effective for specialized applications and in adverse or changeable environments. Engineered for durability and ease of use, the SQ 150 solar panel is manufactured under strict quality controls in ISO 9001 certified facilities.[291]

The panels will be connected in series in banks of ten to produce 240v DC resulting in an electricity production of 1.5KW per bank. The banks will then be linked in parallel configuration to produce the planned total production of 15KW. The configurations will then be arrayed to enable 240v 3 phase AC to be produced. The resultant product is compliant with standard electrical production of Western Power thereby removing the requirement for special transforming and metering devices.

The configuration including connection, inverter and frames is expected to be constructed in situ for around $175,000 for the initial 15 kW, 240v system. Should this system prove effective and viable, a further 100 kW high voltage system is envisaged. The development of which will be separate and distinct from this proposal.

The panels will be linked in banks facing North at an angle of 25° from the vertical such that the surface of the panels will subtend an angle of 90° to the equatorial position of the Sun. The angle will maximize solar collection as winter/summer seasonal variations will range from 22.5° North at the winter solstice (June 26–29) to 22.5° South at the summer solstice (Dec 26–28).[292] Due to the low latitude of the site North–South shading between panels is reduced. Frames will be constructed in such fashion to allow seasonal variations in March and September of each year.

At a latitude of approximately 25° S it is proposed to place the panels at 25° from the horizontal thereby achieving an angle parallel to the tangent of the equator. The north- south seasonal variation of the sun will be from 22.5 o north in winter to 22.5° south in summer. The variation will therefore be no more than 22.5° from 90° to the face of the panels at any time.

---

[291]   Solar/Windup Products Innovation House USA Siemens SP140 solar panels (2002) <http://www.innovationhouse.com/products/solar_siemens_sp140.html> at July 4, 2004.
[292]   Sunrise and sunset for the year 2003 for Latitude 24.880° S Longitude 113.670° E WST provided by Perth Observatory Perth Western Australia 2003.

By fixing the panels horizontal to the east-west axis some loss of production at sunrise and sunset is experienced below 30° above the horizon, as indicated by the experimental results under.[293]

Perpendicular Height (cm)	Angle (o)	Potential Difference (V}
0	0.00	1.50
1	5.31	1.45
2	10.67	1.44
3	16.13	1.43
4	21.74	1.43
5	27.58	1.42
6	33.75	1.40
7	40.40	1.37
8	47.47	1.31
9	56.44	1.18
10	67.81	0.92
10.8	90.00	0.51

Newcastle Upon Tyne Royal Grammar School. *Panel Angle Table of findings*

Evidence gathered from the proponent's 1.92kW plant in Carnarvon demonstrates that daily production is between 3.5 times plant capacity in winter and 5.5 times plant capacity in summer. Observations of production suggest levels can be enhanced by bias to the West as Carnarvon experiences significant light cloud in the early morning with clear skies from around 10 am to sunset. The current plant is biased in an Easterly direction which exacerbates the problem of morning cloud. Further the current array is fixed preventing seasonal adjustment thus exacerbating the variation between summer and winter production. Seasonal adjustment will increase production performance for both summer and winter.

Whilst it is recognised that tracking devices would ensure maximum output of the solar panels during daylight, the complexities of maintenance and spacing between arrays to prevent shading as well as cost, is outweighed simply by enabling more panels to be fixed to the site. Recently developed computer technology provides for rectification and control of power produced to be monitored and controlled such that the panels will produce a constant flow over current regardless of the variation of the angle of the sun's rays to the panels, within certain limitations.

---

[293] Newcastle Upon Tyne Royal Grammar School *Panel Angle* <http://www.rgs.newcastle.sch.uklns ep/Solar%20Energy/angle/angle.html>.

Typical arrangement of the proposed static arrays.

Advice from engineers at Western Power is that production of 15kW, 240v is preferred at this stage to minimise impact on existing electrical demands and infrastructure.

This production is also preferred between 1100h and 1400h when maximum demand is placed on the electrical utility. It is during those times that the solar farm is at peak production at nearly 100% of its capacity. In order to produce 15kW only **100** panels are required[294] the site therefore has ample space for the arrays and incidental infrastructure.

A pilot plant of 1.92kW has been constructed near the site to accurately ascertain annual productions and monitored over six months. The product is consumed domestically and the balance distributed to Western Power at a price of 12.87c per kW in accordance with Western Power's existing REBS purchase system.

Cost of Construction

The solar panels are available from the Advanced Energy Systems at $ 850 plus shipping and engineering supervision as necessary.

In addition to the panels, fencing, framing, a 15 kva transformer/inverter/controller as well as metering devices and feeder connectors are required. The land will require some earthworks and specialized trades people for installation. Government and other authority zoning and planning approvals and fees will also add to the cost of construction.

It is proposed at this stage that the following budget be considered.

---

[294]   Assume electrical production to be at 90% of full capacity due to angle incidence; therefore, 140 W per panel.

Panels 100 @ $850[295]	$85,000
Freight, Insurance[296]	$6,000
Transformer/Inverter/Controller	$20,000
Metering and connection[297]	$10,000
Land clearing and earthworks	$2,000
Fencing 300m @ $30m	$9,000
Framing[298]	$12,500
Incidentals, Engineering, Erection Labour	$10,000
Project Management	$20,000
Total estimated Construction Cost	**$174,500**

Annual Production

Estimated Annual Production based on trial performance of pilot plant.

24,230kW@ $0.12	$2,907
Greenhouse credits 24@ $40[299]	$960
Estimated annual revenue	$3.867

2.2%pa of capital investment.

Funding

It is proposed to finance the development of the land and construction of the infrastructure by way of borrowings from the Bank of Western Australia secured by way of mortgage security over the remainder of the unsecured real estate assets of A R & J Fullarton including Lot 42 Boor Street and a floating debenture over the assets of the company being the infrastructure and contracts for the sale of power of the proposed solar farm. Additional funding to be attracted by way of redeemable preference shares to private venturers, grants, and subsidies for the development of sustainable energy from State and Federal bodies.

The development of environmentally friendly electrical energy production is viewed as a most favorable substitute to the use of fossil fuels for electrical generation. It with this ongoing environmental consideration as a key aspect that financial institutional, government and long term private investment will be solicited. This proposal forms part of those financial support approaches and is pivotal upon them.

Government Sustainable Energy Incentive Grant (55%)	$95,975
Proponent's Investment	$30,525
Bank of Western Australia	$48,000
	$174,500

---

[295] GST is not a consideration as it will either not be imposed or is refundable.
[296] Delivery to Carnarvon via Road transport.
[297] As per quotation by AES attached.
[298] Estimated at $1250 per bank.
[299] Based on Australian Greenhouse Office report price for 2005 available at <http://www.green house.gov.au/markets/mret/mmalpubs/mma_report.pdf> at 3 July 2004.

Service of Loans

$7,296 annual repayments.[300] Shortfall funded by proponent principal repayment averaging $ 4,800 pa.

Cost of Operation

Ongoing fuel costs are nil, that is the purpose of the installation. Routine maintenance is required to wash the solar panels and to inspect for corrosion. The only working mechanisms are the transformers and inverters. Contracted electrical routine maintenance estimated to be negligible. Washing of panels and monitoring to be carried out by the proponents.

Total Annual Outgoings (includes principal repayment of $ 3,100)	$7,296
Total Annual Income	$3,867
Interest Rates & Taxes	$5,200
Net Annual Loss to owners	$1,333
Income Tax @ 30% after depn.	Nil

Licensing, Planning and Insurance

The Shire of Carnarvon have been advised of this proposal and have given tacit support although formal application and planning approval have not been officially applied. Costs associated with these expenses have not been specifically identified.

Insurance

Investigations of Elders Ltd reveal that no insurance is available for renewable energy systems as no such underwriter exists in Australia at this time. Extensive research was recently carried out by Elders Insurance recently to underwrite the electricity project at Exmouth. Despite extensive investigation by one of Australia's rural insurers no underwriter could be procured. The limited size of this plant and the existence of advanced power tracking, transforming and rectifying technology should render damages caused directly by this plant to non-existent levels. Due to the controlling and tracking systems the current flow on the grid controls the production quality. Current loss on the grid for deliberate or accidental reasons immediately switches the plant off

Sudden and Catastrophic events

It is proposed that the solar farm be constructed of only 15kW in the first instance to provide electricity to the 240v system in the Grey's Plain area it is of comparatively

---

[300] Principal and interest repayments over ten years at 9%.

small significance that sudden and catastrophic events will have little impact on Western Powers existing system and current reserve capacity.

Seasonal and daily variations of production will generally coincide with seasonal and daily variations of demand. Peak demand is generally in high summer with the town usage of air-conditioning units and daily demand is generally in the middle of the day.

The Western Power infrastructure is more than adequate to cope with cooler periods and night time consumption. The prime directive of the solar farm is to supplement the existing gas fired plant not replace it. The key saving is reducing, if not entirely preventing reliance on additional electrical production through the use of oil fired engines.

Ongoing technical and engineering support

Has yet to be finalized, however negotiations have been opened with AES Ltd, a Western Australian Publicly Listed company which supplies Shell Solar products and manufactures and exports solar infrastructure. Further information on this company can be found on the worldwide web at www.aesltd.com au

Loss or damage

The proponents are aware that should loss or damage occur to the plant they will bear that loss, subject rights at common law on third parties responsible for that loss including Western Power if that third party is proven to have caused the loss or damage by any means whatsoever and to recover such damages as afforded by Law. WPC will not be expected to purchase the plant should the proponents wish to terminate the project for any reason in the future.

Metering

It is proposed that an export meter of WPC standard be provided at the proponents cost. The meter to be secured appropriately at the entrance to Lot 42 in Boor Street and read in accordance with the standard billing cycle for that area. Placement of the meter will be of greater access than any of the properties in that locality, thereby providing easy access to meter readers. On average production should be around 4,000 KW in a standard billing cycle of 60 days.

Solar Radiation

Engineering advice is that the higher the radiation and the lower the panel temperature, the more efficient the production. Lot 42 is a mere 2 km from the sea over flat ground which provides no impedance to daily sea breezes. BOM data provides that Carnarvon receives over 600Nm of wind per day. The recording centre is no more than 2 km from Lot 42.

This innovative, beneficial and environmental responsible project from which the entire community of Carnarvon and all Western Australians will benefit is responsible and highly exciting. Though early financial returns are limited it is viable in the long term. Carnarvon's location and climate provides a relatively safe environment for the exploitation of solar energy.[301]

The only natural danger is from occasional tropical rotating storms which Carnarvon endures on odd occasion but is not in the "cyclone path" which exists from Broome to Exmouth. Structures will be constructed with this threat in consideration. Carnarvon has never endured a hail storm which is the major threat to solar farms. No fall of hail has ever been recorded in Carnarvon. It is strongly recommended to all parties to the project.

World Performance Solar Radiation Distribution[302]

Carnarvon is denoted +

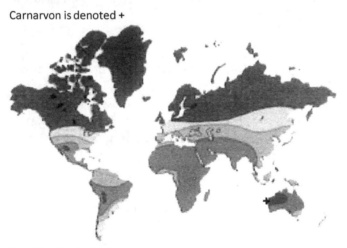

Zone 4; Zone 3; Zone 2; Zone 1

---

[301] Siemens *World Zones for Daily Radiation Performance Information* available at <http://www. oksolar. com/technical/ daiy_solar_radiation.html>.

[302] Distribution chart available at <http://www. oksolar.com/technical/ daiy_solar_radiation. html>.

APPENDIX B

Your Ref:

Our Ref: Proposal for 2MW Solar Farm March 2003 Document OMS#: 1448023

Western Power

Enquiries: Kim Varvell
Telephone: 9441 3470

08 May 2003
Mr Lex Fullarton
A.R. Fullarton & Associates
PO Box 959
CARNARVON WA 6701

Dear Lex

ECONOMIC EVALUATION OF YOUR PROPOSAL

Thank you for forwarding a copy of your proposal for a 2MW solar power station at Carnarvon and the associated financial evaluation in soft copy form.

Western Power has reworked your projections based on confirmation of benchmark costs for the implementation of solar farms, economic indicators used for Western Power's financial projections, marginal cost of displacing generation in Carnarvon, the generation acceptable to Western Power and other operational matters. This has been included in a table for your information at Appendix 1.

Our evaluation shows a Net Present Value of negative $A14.275 M over the 25-year period for a 2 MW solar farm. Our assumptions are stated in the table. The most significant differences are:

- The benchmark cost for installed solar farm capacity is $A1,100 per KW, a total of $A22M for 2 MW;
- The approximate unit cost Western Power will pay for embedded generation is 8.5c/KWH. This figure is not the final figure.
- The level of production from the panels. Although Western Power will take up to 6.3 GWH per annum, it is more likely the panels will generate the figures shown in the table.

Please advise whether you intend to continue with this proposal. If you do continue, there may be additional costs to ensure system stability with the connection of the solar farm. Western Power has not conducted these studies at this time.

Yours sincerely

KIMVARVELL COMMERCIAL MANAGER REGIONAL BRANCH

Western Power Corporation
Regional Branch 510 Abernethy Road Kewdale WA 6105 PO Box 79 Cloverdale WA 6105
Telephone (08) 9441 3400 Facsimile (08) 9441 3409 http://www.w corp.com.au UA3000

Appendix 1 - WPC Financial Evaluation 2MW Solar Farm for Carnarvon							
		Yr0	Yr1	Yr2	Yr3	Yr4	Yr5
Economic Indicators	Base						
CPI Inflator	2004		142.8	146.5	150.5	154.4	158.7
Interest Rate (access Economics)			5.7%	6.0%	6.0%	6.3%	6.2%
Base Tariff							
Escalated Tariff 50% of CPI	$0.085		$0.0852	$0.0852	$0.0852	$0.0852	$0.0852
Production							
Capacity (kW)	2000						
Maximum Output per Day (kWh)	19680						
Per Year x No Days (kWh)	320		6297600				
Conversion Efficiency Losses	95%						
Aging Efficiency per annum	99.2%						
Annual MWH Available			5982.72	5934.86	5887.38	5840.28	5793.56
Production WPC can take	6300						
REC's							
Base Recs	$35.00						
Escalated REC's 80% of CPI			$35.00	$35.73	$36.50	$37.27	$38.12
No. of Recs			5983	5935	5887	5840	5794
Capital Costs							
Installed Capacity/Connection	2000						
Cost per kW	$11,000	$22,000,000					
Funding							
Displaced Diesel Generation Total							
SEDO Grant at 55%		$0					
Loan/Equity		$22,000,000					
Total Capital		$22,000,000					
Capital Costs							
Principal Repayment (Straight Line)	$880,000		$880,000	$880,000	$880,000	$880,000	$880,000
Principal			$21,120,000	$20,240,000	$19,360,000	$18,480,000	$17,600,000
Interest Payable			$1,203,840	$1,214,400	$1,161,600	$1,164,240	$1,091,200
Revenue							
Production			$509,896	$505,817	$501,770	$497,756	$493,774
REC Revenue			$209,395	$212,041	$214,911	$217,672	$220,841
Total Revenue			**$719,291**	**$717,858**	**$716,681**	**$715,428**	**$714,615**
Costs							
Lease Fees	$30,000.00		$30,000.00	$30,624.00	$31,288.80	$31,946.40	$32,672.90
Provision for Maintenance	$200,000.00		$200,000.00	$204,160.00	$208,592.30	$212,975.70	$217,819.00
Loan Interest			$1,203,840	$1,214,400	$1,161,600	$1,164,240	$1,091,200
Redeemable Shares			$70,000	$70,000	$70,000	$70,000	$70,000
Total Costs			**$1,503,840**	**$1,519,184**	**$1,471,481**	**$1,479,162**	**$1,411,692**
Net Profit			-$784,549	-$801,326	-$754,800	-$763,734	-$697,077
Depreciation Straight Line			$880,000	$880,000	$880,000	$880,000	$880,000
Assessable Income			-$1,664,549	-$1,681,326	-$1,634,800	-$1,643,734	-$1,577,077
Taxation @ 30%			-$499,365	-$504,398	-$490,440	-$493,120	-$473,123
Principal Repayment			$880,000	$880,000	$880,000	$880,000	$880,000
Distributable Funds			-$1,165,184	-$1,176,928	-$1,144,360	-$1,150,614	-$1,103,954
Cashflow							
Revenue		$22,000,000	$719,291	$717,858	$716,681	$715,428	$714,615
Costs		$22,000,000	$2,883,205	$2,903,582	$2,841,921	$2,852,282	$2,764,815
Cashflow		$0	-$2,163,913	-$2,185,724	-$2,125,240	-$2,136,854	-$2,050,200
Cumulative Cashflow		$0	-$2,163,913	-$4,349,637	-$6,474,877	-$8,611,731	-$10,661,931
NPV		-$14,274,924					

Appendix 1 - WPC Financial Evaluation 2MW Solar Farm for Carnarvon cont:-					
	Yr6	Yr7	Yr8	Yr9	Yr10
Economic Indicators					
CPI Inflator	163.8	168.7	173.3	178	183.1
Interest Rate (access Economics)	6.4%	6.5%	6.5%	6.5%	6.5%
Base Tariff					
Escalated Tariff 50% of CPI	$0.0915	$0.0915	$0.0915	$0.0915	$0.0915
Production					
Capacity (kW)					
Maximum Output per Day (kWh)					
Per Year x No Days (kWh)					
Conversion Efficiency Losses					
Aging Efficiency per annum					
Annual MWH Available	5747.21	5701.23	5655.62	5610.38	5565.49
Production WPC can take					
REC's					
Base Recs					
Escalated REC's 80% of CPI	$39.11	$40.08	$40.97	$41.89	$42.90
No. of Recs	5747	5701	5656	5610	5565
Capital Costs					
Installed Capacity/Connection					
Cost per kW					
Funding					
Displaced Diesel Generation Total					
SEDO Grant at 55%					
Loan/Equity					
Total Capital					
Capital Costs					
Principal Repayment (Straight Line)	$880,000	$880,000	$880,000	$880,000	$880,000
Principal	$16,720,000	$15,840,000	$14,960,000	$14,080,000	$13,200,000
Interest Payable	$1,070,080	$1,029,600	$972,400	$915,200	$858,000
Revenue					
Production	$525,810	$521,603	$517,430	$513,291	$509,185
REC Revenue	$224,797	$228,491	$231,714	$235,007	$238,757
Total Revenue	**$750,607**	**$750,094**	**$749,144**	**$748,298**	**$747,942**
Costs					
Lease Fees	$33,526.40	$34,352.20	$35,117.70	$35,903.90	$36,771.10
Provision for Maintenance	$223,509.20	$229,014.50	$234,117.90	$239,359.10	$245,140.50
Loan Interest	$1,070,080	$1,029,600	$972,400	$915,200	$858,000
Redeemable Shares					
Total Costs	**$1,327,116**	**$1,292,967**	**$1,241,636**	**$1,190,463**	**$1,139,912**
Net Profit	-$576,509	-$542,872	-$492,491	-$442,165	-$391,970
Depreciation Straight Line	$880,000	$880,000	$880,000	$880,000	$880,000
Assessable Income	-$1,456,509	-$1,422,872	-$1,372,491	-$1,322,165	-$1,271,970
Taxation @ 30%	-$436,953	-$426,862	-$411,747	-$396,649	-$381,591
Principal Repayment	$880,000	$880,000	$880,000	$880,000	$880,000
Distributable Funds	-$1,019,556	-$996,011	-$960,744	-$925,515	-$890,379
Cashflow					
Revenue	$750,607	$750,094	$749,144	$748,298	$747,942
Costs	$2,644,068	$2,599,828	$2,533,383	$2,467,112	$2,401,503
Cashflow	-$1,893,461	-$1,849,734	-$1,784,238	-$1,718,814	-$1,653,561
Cumulative Cashflow	-$12,555,394	-$14,405,128	-$16,189,367	-$17,908,180	-$19,561,742
NPV					

Appendix 1 - WPC Financial Evaluation 2MW Solar Farm for Carnarvon cont:-					
	Yr11	Yr12	Yr13	Yr14	Yr15
Economic Indicators					
CPI Inflator	188.6	194.5	200.7	206.9	213.1
Interest Rate (access Economics)	6.5%	6.5%	6.5%	6.5%	6.5%
Base Tariff					
Escalated Tariff 50% of CPI	$0.0989	$0.0989	$0.0989	$0.0989	$0.0989
Production					
Capacity (kW)					
Maximum Output per Day (kWh)					
Per Year x No Days (kWh)	6297600				
Conversion Efficiency Losses					
Aging Efficiency per annum					
Annual MWH Available	5520.97	5476.80	5432.99	5389.52	5346.41
Production WPC can take					
REC's					
Base Recs					
Escalated REC's 80% of CPI	$43.98	$45.12	$46.34	$47.56	$48.78
No. of Recs	5521	5477	5433	5390	5346
Capital Costs					
Installed Capacity/Connection					
Cost per kW					
Funding					
Displaced Diesel Generation Total					
SEDO Grant at 55%					
Loan/Equity					
Total Capital					
Capital Costs					
Principal Repayment (Straight Line)	$880,000	$880,000	$880,000	$880,000	$880,000
Principal	$12,320,000	$11,440,000	$10,560,000	$9,680,000	$8,800,000
Interest Payable	$800,800	$743,600	$686,400	$629,200	$572,000
Revenue					
Production	$545,968	$541,600	$537,267	$532,969	$528,706
REC Revenue	$242,793	$247,129	$251,780	$256,339	$260,794
Total Revenue	**$788,761**	**$788,729**	**$789,047**	**$789,308**	**$789,500**
Costs					
Lease Fees	$37,694.20	$38,676.70	$39,722.40	$40,767.80	$41,810.80
Provision for Maintenance	$251,294.70	$257,844.90	$264,815.90	$271,785.20	$278,738.80
Loan Interest	$800,800	$743,600	$686,400	$629,200	$572,000
Redeemable Shares					
Total Costs	**$1,089,789**	**$1,040,122**	**$990,938**	**$941,753**	**$892,550**
Net Profit	-$301,028	-$251,393	-$201,891	-$152,445	-$103,050
Depreciation Straight Line	$880,000	$880,000	$880,000	$880,000	$880,000
Assessable Income	-$1,181,028	-$1,131,393	-$1,081,891	-$1,032,445	-$983,050
Taxation @ 30%	-$354,308	-$339,418	-$324,567	-$309,733	-$294,915
Principal Repayment	$880,000	$880,000	$880,000	$880,000	$880,000
Distributable Funds	-$826,719	-$791,975	-$757,324	-$722,711	-$688,135
Cashflow					
Revenue	$788,761	$788,729	$789,047	$789,308	$789,500
Costs	$2,324,097	$2,259,539	$2,195,506	$2,131,486	$2,067,465
Cashflow	-$1,535,336	-$1,470,810	-$1,406,459	-$1,342,178	-$1,277,965
Cumulative Cashflow	-$21,097,077	-$22,567,888	-$23,974,347	-$25,316,524	-$26,594,489
NPV					

Appendix 1 - WPC Financial Evaluation 2MW Solar Farm for Carnarvon cont:-					
	Yr16	Yr17	Yr18	Yr19	Yr20
Economic Indicators					
CPI Inflator	219.5	225.9	232.4	238.9	245.6
Interest Rate (access Economics)	6.5%	6.5%	6.5%	6.5%	6.5%
Base Tariff					
Escalated Tariff 50% of CPI	$0.1081	$0.1081	$0.1081	$0.1081	$0.1081
Production					
Capacity (kW)					
Maximum Output per Day (kWh)					
Per Year x No Days (kWh)					
Conversion Efficiency Losses					
Aging Efficiency per annum					
Annual MWH Available	5303.64	5261.21	5219.12	5177.37	5135.95
Production WPC can take					
REC's					
Base Recs					
Escalated REC's 80% of CPI	$50.03	$51.28	$52.56	$53.84	$55.15
No. of Recs	5304	5261	5219	5177	5136
Capital Costs					
Installed Capacity/Connection					
Cost per kW					
Funding					
Displaced Diesel Generation Total					
SEDO Grant at 55%					
Loan/Equity					
Total Capital					
Capital Costs					
Principal Repayment (Straight Line)	$880,000	$880,000	$880,000	$880,000	$880,000
Principal	$7,920,000	$7,040,000	$6,160,000	$5,280,000	$4,400,000
Interest Payable	$514,800	$457,600	$400,400	$343,200	$286,000
Revenue					
Production	$573,359	$568,772	$564,222	$559,708	$55,231
REC Revenue	$265,355	$269,798	$274,342	$278,752	$283,258
Total Revenue	**$838,714**	**$838,570**	**$838,564**	**$838,461**	**$338,489**
Costs					
Lease Fees	$42,885.10	$43,954.80	$45,055.50	$46,149.00	$47,273.20
Provision for Maintenance	$285,900.90	$293,032.00	$300,370.00	$307,660.30	$315,154.80
Loan Interest	$514,800	$457,600	$400,400	$343,200	$286,000
Redeemable Shares					
Total Costs	**$843,586**	**$794,587**	**$745,825**	**$697,009**	**$648,428**
Net Profit	-$4,872	$43,983	$92,738	$141,451	$190,061
Depreciation Straight Line	$880,000	$880,000	$880,000	$880,000	$880,000
Assessable Income	-$884,872	-$836,017	-$787,262	-$738,549	-$689,939
Taxation @ 30%	-$265,462	-$250,805	-$236,179	-$221,565	-$206,982
Principal Repayment	$880,000	$880,000	$880,000	$880,000	$880,000
Distributable Funds	-$619,410	-$585,212	-$551,083	-$516,984	-$482,957
Cashflow					
Revenue	$838,714	$838,570	$838,564	$838,461	$838,489
Costs	$1,989,048	$1,925,392	$1,862,004	$1,798,574	$1,735,410
Cashflow	-$1,150,333	-$1,086,822	-$1,023,440	-$960,113	-$896,921
Cumulative Cashflow	-$27,744,823	-$28,831,644	-$29,855,084	-$30,815,198	-$31,712,118
NPV					

Appendix 1 - WPC Financial Evaluation 2MW Solar Farm for Carnarvon cont:-	Yr21	Yr22	Yr23	Yr24	Yr25
Economic Indicators					
CPI Inflator	252.5	259.3	266.1	273	279.8
Interest Rate (access Economics)	6.5%	6.5%	6.5%	6.5%	6.5%
Base Tariff					
Escalated Tariff 50% of CPI	$0.1180	$0.1180	$0.1180	$0.1180	$0.1180
Production					
Capacity (kW)					
Maximum Output per Day (kWh)					
Per Year x No Days (kWh)					
Conversion Efficiency Losses					
Aging Efficiency per annum					
Annual MWH Available	5094.86	5054.10	5013.67	4973.56	4933.77
Production WPC can take					
REC's					
Base Recs					
Escalated REC's 80% of CPI	$56.50	$57.84	$59.17	$60.51	$61.85
No. of Recs	5095	5054	5014	4974	4934
Capital Costs					
Installed Capacity/Connection					
Cost per kW					
Funding					
Displaced Diesel Generation Total					
SEDO Grant at 55%					
Loan/Equity					
Total Capital					
Capital Costs					
Principal Repayment (Straight Line)	$880,000	$880,000	$880,000	$880,000	$880,000
Principal	$3,520,000	$2,640,000	$1,760,000	$880,000	$0
Interest Payable	$228,800	$171,600	$114,400	$57,200	$0
Revenue					
Production	$600,940	$596,132	$591,363	$586,632	$581,939
REC Revenue	$287,861	$292,313	$296,676	$300,949	$305,136
Total Revenue	**$888,801**	**$888,445**	**$888,039**	**$887,581**	**$887,075**
Costs					
Lease Fees	$48,428.90	$49,574.50	$50,720.00	51865.6	53011.2
Provision for Maintenance	$322,859.20	$330,496.40	$338,133.60	345770.7	353407.9
Loan Interest	$228,800	$171,600	$114,400	$57,200	$0
Redeemable Shares					
Total Costs	**$600,088**	**$551,671**	**$503,254**	**$454,836**	**$406,419**
Net Profit	$288,713	$336,774	$384,785	$432,745	$480,656
Depreciation Straight Line	$880,000	$880,000	$880,000	$880,000	$880,000
Assessable Income	-$591,287	-$543,226	-$495,215	-$447,255	-$399,344
Taxation @ 30%	-$177,386	-$162,968	-$148,565	-$134,176	-$119,803
Principal Repayment	$880,000	$880,000	$880,000	$880,000	$880,000
Distributable Funds	-$413,901	-$380,258	-$346,651	-$313,078	-$279,541
Cashflow					
Revenue	$888,801	$888,445	$888,039	$887,581	$887,075
Costs	$1,657,474	$1,594,639	$1,531,618	$1,469,013	$1,406,222
Cashflow	-$768,673	-$706,193	-$643,780	-$581,431	-$519,148
Cumulative Cashflow	-$32,480,792	-$33,186,985	-$33,830,765	-$34,412,196	-$34,931,344
NPV					

INDEX

BIBLIOGRAPHY

ARTICLES/BOOKS/REPORTS

Alexander, Gary, and Boyle, Godfrey, "Introducing Renewable Energy," in Godfrey Boyle (ed) *Renewable Energy: Power for a Sustainable Future* (2nd ed, 2004).

Arceivala, Soli J, and Asolekar, Shyam R, *Environmental Studies: A Practitioner's Approach* (2012).

Arnalich, Santiago, *Epanet and Development. How to Calculate Water Networks by Computer* (2011).

Australian Energy Market Commission, *2014 Residential Electricity Price Trends Report* (2014).

Brundtland, Gro Harlem, *Report of the World Commission on Environment and Development: Our Common Future* UN Doc a/42/427 (1987).

Boyle, Godfrey, "Solar Photovoltaics," in Godfrey Boyle (ed) *Renewable Energy: Power for a Sustainable Future* (2nd ed, 2004), p. 66.

CCH Australia Ltd, *The Asprey Report: An Analysis* (1975).

Chint Solar (Zhejiang), Co., Ltd, *Datasheet Crystalline PV Modules CHSM 6610P series Astronergy 235wp* (2012).

Emergency Management Australia, *Hazards, Disasters and Survival: A Booklet for Students and the Community* (1997).

Fitzpatrick, Merton George John, *Daurie Creek* (2004).

Fullarton, Alexander Robert, *Heat, Dust and Taxes* (2015).

GE Energy, *GEPV-110 110 Watt Photovoltaic Module* (2004).

Government of Western Australia, *Hope for the Future: The Western Australian State Sustainability Strategy: A Vision for Quality of Life in Western Australia* (2003).

Houghton, Sir John Theodore, *Global Warming: The Complete Briefing* (3rd ed, 2004).

Iqbal, Muhammad, *An Introduction to Solar Radiation* (1983).

Lewis, Simon, Carnarvon: *A Case Study of Increasing Levels of PV Penetration in an Isolated Electricity Supply System* (Report prepared by The Centre for Energy and Environmental Markets, 2012).

Milne, Grace Veronica "Bonnie," *Amelia* (2002).

Milne, Grace Veronica "Bonnie," Crossroads: *A Journey through the Upper Gascoyne* (2007).

Milne, Grace Veronica "Bonnie," *Pioneer Father, Pioneer Son: York to Gascoyne with the Collins Family* (2010).

Mortimore, Anna, "Reforming vehicle taxes on new car purchases can reduce road transport emissions—ex post evidence" (2014), 29 *Australian Tax Forum* 177.

Muir, Ramsay, Bygone *Liverpool illustrated by ninety-seven plates reproduced from original paintings, drawings, manuscripts, and prints with historical descriptions by Henry S. and Harold E. Young* (1913).

Noll, Daniel, Dawes, Colleen, and Rai, Varun, "Solar Community Organizations and active peer effects in the adoption of residential PV" (2014) 67 *Energy Policy* 330–343.

O'Connor, Pamela, "The Private Taking of Land: Adverse Possession, Encroachment by Buildings and Improvement Under a Mistake" (2006) 33 *University of Western Australia Law Review* 31.

Paterson, Andrew Barton (Banjo), *Clancy of the Overflow* (1889).

Passey, Robert, et al, *Study of Grid-connect Photovoltaic Systems—Benefits, Opportunities, Barriers and Strategies: Final Report* (2007).

Passey, Robert, et al, *Study of Grid-connect Photovoltaic Systems: Benefits, Opportunities and Strategies* (2009).

Passey, Robert, et al. "The potential impacts of grid-connected distributed generation and how to address them: A review of technical and non-technical factors" (2011) 39 *Energy Policy* 6280.

Savory, Allan, and Butterfield, Jody, *Holistic Management: A New Framework for Decision Making* (2nd ed 1999).

Sorenson, Jack, *The Lost Shanty* (1939).

Stapleton, Geoff, and Neill, Susan, *Grid-connected Solar Electric Systems: The Earthscan Expert Handbook for Planning Design and Installation* (2012).

Strange, Tracey, and Bayley, Anne, *Sustainable Development: Linking Economy, Society and Environment* (2008).

Sturman, Andrew P, and Tapper, Nigel J, *The Weather and Climate of Australia and New Zealand* (1996).

Sunpower Corporation, *210 Solar Panel 210 Watt Photovoltaic Module* (2009).

Twidell, John, and Weir, Tony, *Renewable Energy Resources* (2nd ed, 2006).

Valli, Jack, *Gascoyne Days* (1983).

Walker Gordon, and Devine-Wright, Patrick, "Community renewable energy: What should it mean" (2008) 36 *Energy Policy* 497–500.

Watts, Duncan J, *Everything is Obvious: How Common Sense Fails* (2011).

Yallop, Colin et al (eds), *Macquarie Concise Dictionary* (4th ed, 2006).

CASE LAW

Fullarton v The Estate of McAllister (Unreported, Supreme Court of Western Australia, Bredmeyer Master, March 17, 1997.

Fullarton v The Estate of McAllister (Unreported, Supreme Court of Western Australia, Sanderson Master, April 17, 2002.

LEGISLATION

Clean Energy Act 2011 (Cth).

Clean Energy Regulations 2011 (Cth).

Environment Protection and Biodiversity Conservation Act 1999 (Cth).

Income Tax Assessment Act 1936 (Cth).

Income Tax Assessment Act 1997 (Cth).

Local Government Act 1995 (WA).

Planning and Development Act 2005 (WA).

Renewable Energy (Electricity) Act 2000 (Cth).

Renewable Energy (Electricity) (Charge) Act 2000 (Cth).

Renewable Energy (Electricity) Bill 2000 (Cth) *Explanatory Memorandum.*

Renewable Energy (Electricity) Regulations 2001 (Cth).

Renewable Energy (Electricity) (Large-scale Generation Shortfall Charge) Act 2000 (Cth).

Renewable Energy (Electricity) (Small-scale Technology Shortfall Charge) Act 2010 (Cth).

OTHER SOURCES

Abbott, Anthony John, "Address to the South Australian Liberal Party Annual General Meeting" (Speech delivered at the South Australian Liberal Party Annual General Meeting, Adelaide, August 15, 2015).

ABC News (Australia), 'Carnarvon residents show solar power can pay', 7.30 Report, April 13, 2013, <https://www.youtube.com/watch?v=1mS1D5yMjLg> at January 8, 2016.

AECOM Australia Pty Ltd, "Australia's Off-Grid Clean Energy Market Research Paper" (Research Paper, Australian Renewable Energy Agency, 2014)

Australian Energy Regulator <https://www.aer.gov.au/> at June 12, 2015.

Australian Energy Regulator, *State of the Energy Market 2010* (2010) <https://www.aer.gov.au/> at June 12, 2015.

Australian Government Bureau of Meteorology <http://www.bom.gov.au/climate/map/temperature/ID CJCM0005 temperature.shtml> at July 19, 2007.

Australian Government Bureau of Meteorology, *Climate Statistics for Australian Locations: Carnarvon Airport* (2014) <http://www.bom.gov.au/climate/averages/tables/cw_006011.shtml> at January 3, 2015.

Australian Government Bureau of Meteorology, *Climate Data Online* (2015) <http://www.bom.g ov.au/climate/data/> at February 9, 2015.

Australian Government Bureau of Meteorology, Carnarvon, Western Australia March 2015 Daily Weather Observations <http://www.bom.gov.au/climate/dwo/IDCJDW6024.latest.shtml > at March 29, 2015.

Australian Government, Clean Energy Regulator, *The History of the Renewable Energy Target* (2014) Renewable Energy Target <http://ret.cleanenergyregulator.gov.au/About-the-scheme/History-of-the-RET> at March 29, 2015.

Australian Government, Clean Energy Regulator, *REC Registry* (2014) < https://www.rec-registry.gov.au/re c-registry/app/public/lgc-register> at March 29, 2015.

Australian Government, Clean Energy Regulator, *Register of Large-Scale Generation Certificates* (2015) <https://www.rec-registry.gov.au/rec-registry/app/public/lgc-register> at June 13, 2015.

Australian Government, Clean Energy Regulator, *The Renewable Power Percentage Target* (2014) Renewable Energy Target <http://ret.cleanenergyregulator.gov.au/About-the-Schemes/About-the-renewable-power-percentage/Annual-targets> at March 29, 2015.

Australian Government, Clean Energy Regulator, *Postcode Data for Small-Scale Installations* (2015) < http://www.cleanenergyregulator.gov.au/RET/Forms-and-resources/Postcode-data-for-small-scale-in-stallations> at July 22, 2015.

Australian Government: Clean Energy Regulator, *Creation of Small-Scale Technology Certificates* (2016) <http://www.cleanenergyregulator.gov.au/About/Accountability-and-reporting/administrative-reports /the-renewable-energy-target-2014-administrative-report/Creation-of-small-scale-technology-certifica tes> at January 13, 2016.

Australian Government, Department of Agriculture, *Australian Energy Statistics—Energy Update 2011* (2015) <http://www.agriculture.gov.au/abares/publications/display?url=http://143.188.17.20/anrdl/DA FFService/display.php?fid=pe_abares99010610_12c.xml> at June 13, 2015.

Australian Government, Department of the Environment, *Australia's Emissions Projections 2014–15* (2015).

Australian PV Institute, *PV Postcode Data* (2015) <http://pv-map.apvi.org.au/analyses> at July 22, 2015.

City of Cockburn, *Sustainability Strategy 2013–2017* (2013).

Clean Energy Council, *Accredited Installer: Accreditation Guidelines* (2014) <http://www.solaraccred itation.com.au/installers/compliance-and-standards/accreditation-guidelines.html> at October 27, 2015.

Clover, Ian, "Western Australia's rooftop solar now state's 'biggest power station'" *Renew Economy* <http://reneweconomy.com.au/2016>, January 7, 2016.

Dairy Australia, *Australian Dairy Shed Energy Costs* (2015) <http://www.dairyaustralia.com.au/~/medi a/Documents/Environment%20and%20Resources/22072014-Australian%20Dairy%20Shed%20Energy% 20Costs-Fact%20Sheet-July14.pdf> at August 8, 2015.

Electricity Networks Corporation trading as Western Power, *Western Power Annual Report 2015* <http://www.westernpower.com.au/documents/annual-report-2015.pdf> at January 10, 2016.

Email from Craig Deetlefs, Horizon Power Corporation to Alexander Robert Fullarton, October 29, 2015 (held by author).

Facsimile from Bureau of Meteorology to AR Fullarton, February 21, 2003.

Facsimile from Perth Observatory to AR Fullarton, February 26, 2003.

Fronius Australia, *Grid-Connect Inverters* (2015) Fronius International GmbH <http://www.fronius.com /cps/rde/xchg/SID-3FCAF372-00042740/fronius_australia/hs.xsl/25_4961.htm#.VhCH9Yahfcc> at October 4, 2015.

Fronius international GmbH, Fronius Australia, Inverter With Multiple MPP Trackers: Requirements and State of the Art Solutions (2013) <http://www.fronius.com/cps/rde/xbcr/sid-1148ef51-e86faada/froni us_international/se_ta_inverter_with_multiple_mpp_trackers_en_320367_snapshot.pdf> at December 13, 2015.

Government of Western Australia, Department of Water, *Water Facts WF13; Flooding in Western Australia* (2000) <https://www.water.wa.gov.au/__data/assets/pdf_file/0014/1670/WER-120-WRCWF13.pdf> at January 3, 2016.

Government of Western Australia: Media Statements, *Feed-in Tariff Scheme Provides Incentive* <https://www.mediastatements.wa.gov.au/Pages/Barnett/2010/05/Feed-in-tariff-scheme-provides-inc entive.aspx> at January 16, 2016.

Government of Western Australia: Media Statements, *Residential Feed-in Tariff Scheme Suspended After Reachinglits Quota* (2011) <https://www.mediastatements.wa.gov.au/Pages/Barnett/2011/08/Residen tial-feed-in-tariff-scheme-suspended-after-reaching-its-quota.aspx> at January 16, 2016.

Halvorsen, Stein, Engineered diagram, "Solex Carnarvon Solar Farm: Solar panel chassis," (2009).

Hybrid Optimisation of Multiple Energy Resources (HOMER) <http://www.homerenergy.com/sof tware.html> at April 27, 2015.

Kalmar, Nicola, "Solar project gets boost from local business," *The West Australian*, <https://au.news.ya-hoo.com/thewest/wa/a/19274310/solar-project-gets-boost-from-local-business/> at October 6, 2013.

Kirkpatrick, David Gordon (Slim Dusty) (Ballad), *The Frog* (1985).

Letter from Ken Baston Western Australian Minister for Agriculture and Food to Vince Catania Western Australian Member for North West Central, January 27, 2015 (copy held by author).

Letter from the Deputy Commissioner of Taxation to Alexander Fullarton, 5 January 2006 (held by author) Notice of Private [Tax] Ruling 59756.

Letter from Grant Stacy, Asset Manager Regional Branch Western Power to Alexander Fullarton, February 18, 2003 (held by author).

Letter from Kim Varvell, Commercial Manager Regional Branch Western Power to Alexander Fullarton, May 8, 2003 (held by author).

Letter from Mike Laughton-Smith, Manager Regional Branch Western Power to Alexander Fullarton, October 21, 2004 (held by author).

Letter from Evan Gray, Senior Program Officer, Western Australian Sustainable Energy Development Office to Alexander Fullarton, April 22, 2004 (held by author).

Mercer, Daniel, "Sun, giant batteries to power Carnarvon," *The West Australian*, 24 November 2015, <https://au.news.yahoo.com/thewest/wa/a/30178788/giant-batteries-to-store-power-geraldton-in-trial/> at January 19, 2016.

Newcastle Upon Tyne Royal Grammar School *Panel Angle* <http://www.rgs.newcastle.sch.ukln sep/Solar%20Energy/angle/angle.html> at July 4, 2004.

Organisation for Economic Co-Operation and Development, *Environment; Air and Climate; Greenhouse Gas Emissions by Source* (2016) <http://stats.oecd.org/Index.aspx?DataSetCode%3DAIR_GHG> at January 9, 2016.

Peters, Jen, *Solar Energy* (2003) University of Wisconson-Eau Claire <http://academic.evergree n.edu/g/grossmaz/petersj.html> at April 19, 2015.

Photograph taken by Alexander R Fullarton (1994).

Photograph taken by Alexander R Fullarton (2004).

Photograph taken by Alexander R Fullarton (2008).

Photograph taken by Alexander R Fullarton (2009).

Photograph taken by Alexander R Fullarton (2010).

Photograph taken by Alexander R Fullarton (2013).

Photograph taken by Alexander R Fullarton (2015) (1).

Photograph taken by Alexander R Fullarton (2015) (2).

Photograph courtesy of Mitsubishi Motors Australia Facebook Page (2014) Facebook <https://www.fac ebook.com/MitsubishiMotorsAustralia/?hc_location=ufi> at January 8, 2016.

Regional Power Corporation trading as Horizon Power, *Annual Report 2006* (2006).

Regional Power Corporation trading as Horizon Power, *Financial Statements for the Year Ended 30 June 2015* (2015).

Regional Power Corporation trading as Horizon Power, *Reports and Publications: Annual Reports* (2016) <http://horizonpower.com.au/about-us/overview/reports-publications/annual-reports/> at January 2, 2016.

Regional Power Corporation trading as Horizon Power, *Marble Bar and Nullagine Solar Power Stations* (2016) <http://horizonpower.com.au/about-us/our-assets/marble-bar-and-nullagine-solar-power-statio ns/> at January 10, 2016.

Regional Power Corporation trading as Horizon Power, *Understanding Renewable Energy*, (2016) <http://horizonpower.com.au/being-energy-efficient/solar/understanding-renewable-energy/> at January 15, 2016.

ROAM Consulting Pty Ltd, "Solar Generation Australian Market Modelling" (Report to the Australian Solar Institute, 2012).

Smith, Sean, "EMC Solar to open Carnarvon Project," *The West Australian*, May 4, 2012, <https://au.news.ya-hoo.com/thewest/wa/a/13599005/emc-solar-to-open-carnarvon-project/> at January 19, 2016.

Solex data held by Alexander Fullarton.

Sonti, Chalpat, "Solar power station a step closer to reality," *The Sydney Morning Herald*, February 16, 2010, <http://www.smh.com.au/environment/energy-smart/solar-power-station-a-step-closer-to-realit y-20100216-o6in.html> January 15, 2016.

The Greens <http://greens.org.au/our-story> at February 13, 2015.

US Energy Information Administration *International Energy Statistics* (2012) <http://www.eia.go v/cfapps/ipdbproject/iedin-dex3.cfm?tid=6&pid=29&aid=12&cid=CG5,&syid=2008&eyid=2012&unit=BKWH> at August 29, 2015.

University of Birmingham, *BiSON Background: An Overview* (1025) University of Birmingham <http://bi-son.ph.bham.ac.uk/index.php?page=bison,background> at April 20, 2015.

Vorrath, Sophie, "Dubbo mayor drives fully electric Nissan Leaf, powered by the sun" *Renew Economy* <http://reneweconomy.com.au/2016> at January 25, 2016.

Western Australia Certificate of Title Volume 664 Folio 74.

Western Australian Government, Department of Finance, *Electricity Pricing* (2015) <http://www.fi-nance.wa.gov.au/cms/content.aspx?id=15096> at September 29, 2015.

Western Power, Reports and Publications: Annual Reports (2016) <http://www.westernpower.com.au/cor porate-information-annual-reports.html> and <http://www.parliament.wa.gov.au/publications/t abledpapers.nsf/displaypaper/3621468a5f3bae3bbab5a64f48256d9c00323e96/$file/western+power+c orporation+ar+2003.pdf> at January 2, 2016

ibidem-Verlag / *ibidem* Press
Melchiorstr. 15
70439 Stuttgart
Germany

ibidem@ibidem.eu
ibidem.eu

Printed in the USA
CPSIA information can be obtained
at www.ICGtesting.com
JSHW011517221024
72172JS00006B/51

9 783838 208640